WHAT COLOR IS A CONSERVATIVE?

WHAT COLOR IS A CONSERVATIVE?

My Life and My Politics

J. C. Watts, Jr.

with Chriss Winston

HarperCollins*Publishers*

HarperCollins books may be purchased for educational, business, or sales promotional use. For information, please write: Special Markets Department, HarperCollins Publishers Inc., 10 East 53rd Street, New York, NY 10022.

FIRST EDITION

Designed by Nancy Singer Olaguera

Printed on acid-free paper

Library of Congress Cataloging-in-Publication Data
Watts, J. C. (Julius Caesar).
 What color is a conservative? : my life and my politics / by J. C. Watts, Jr., with Chriss Winston.—1st ed.
  p. cm.
 Includes index.
 ISBN 0-06-019436-7 (acid-free paper)
 1. Watts, J. C. (Julius Caesar), date. 2. African American legislators—Biography. 3. Legislators—United States—Biography. 4. United States. Congress. House—Biography. 5. Conservatism—United States—History—20th century. 6. African Americans—Politics and government—20th century. 7. United States—Politics and government—1989– I. Winston, Chriss Anne, date. II. Title.
E840.8.W39 A3 2002
328.73'092—dc21
                                                        2002023271

03 04 05 06 ❖/RRD 10 9 8 7 6 5

To Frankie, who has loved me and has been a Rock of Gibraltar through thick and thin;

To my kids, LaKesha, Jerrell, Jennifer, Julia, Trey, and Tia, for not beating up on me for all your events I've had to miss;

To my brothers and sisters for being such great role models when it came to the work ethic; and

To my late parents for teaching us that work ethic.

CONTENTS

INTRODUCTION

> There is never a beginning, there is never an end, to the inexplicable continuity of this web of God, but always circular power returning into itself.
>
> Ralph Waldo Emerson

On Monday, July 1, 2002, at ten A.M., I called a press conference, one of the more significant ones of my life. A number of conversations had led up to it—with President Bush, Vice President Cheney, my constituents, my staff, my extended family, my wife, Frankie, and many others from both sides of the party lines. I also received a powerful letter from Rosa Parks, who urged me to stay in office, to "remain as a pioneer on the Republican side until others come to assist you. . . . I would also like you to keep your seat, and not think of your mantel as heavy, but think that you are chosen to prepare the way."

They made my decision much more difficult, but after eight years of serving the 4th District of Oklahoma, I was convinced that it was time for me to step down and again become a citizen. President Bush understood my reasons. He is a proud father himself and knows the importance of my spending a little more time with my family. I also explained to him that with his election and the leadership he was providing, I felt more comfortable about my decision. Vice President Cheney, who had been through a similar situation midterm in his public service, was very understanding. Many others were stunned that I had decided not to run for reelection, but I had to follow my heart and my conscience, which I have always tried to do.

For eight years I served in Congress, and it provided me with an enormous amount of satisfaction and a sense of accomplishment. It has been a great honor and I've experienced some of the most exhilarating moments of my life. I have worked with great people, some of the best in the world of politics. Certainly, one of the most difficult moments in my office on the Hill was looking at each of them—Pam Pryor, Tom Cole, Mike Hunter, Jon Vandenheuvel, Elroy Sailor, Congressman Hoekstra and others whom I had come to depend on—and explaining that this session of Congress would be my last, that it was time for me to go home. Their expressions displayed a wide spectrum of feelings—acceptance, surprise, disbelief—and I'll admit I had a hard time holding myself together. We had been side by side in the trenches, had believed in each other, and, come January, we could go our separate ways. We have so much to be proud of: reducing taxes not once but twice, transforming the welfare system, starting down that long part of rebuilding the military, the passage of the Community Renewal Act, the faith-based initiative, and much, much more. It has been a wonderful ride.

Alongside me through all this has been my patient, devoted, and loving wife, Frankie. She has been nothing short of magnificent, supporting me as I would spend the week in D.C., then fly down for the weekend to be with her and our three youngest children in Norman, Oklahoma. Through the late sessions and the travel, she has managed the house and the children without complaint—and, most of the time, without me. While I do speak to her every day and touch base with the kids each night, I am feeling the tug of wanting to be with my family more, to do things that others take for granted more often—dinner where we can talk about school and work, playing catch in the yard, helping with homework, or simply sitting on the porch and growing old with the woman I love. I have asked a lot of Frankie and my family over the years, and now I want to just *be* with them more.

It doesn't mean I'm going to up and disappear from public discourse, though. I care about the greatest country on earth too much not to actively debate about the important issues that will affect America in the years ahead. (More on that in chapter 13.) Aside from the fact that it's in my blood, it is also my responsibility as a citizen of a democracy, a responsibility that I take seriously.

Of course, things being the way they are, my announcement spurred more speculation than the futures market, with headlines like "How His Party Frustrated Watts," and questions like "What do you think it says about the Republican Party that the only Republican African-American serving in the Congress is choosing to step down?" Amazing. Still, I understand the cynicism, an attitude that is not exactly lacking in the media. It is difficult for some to process how a politician is willing to give up his power while so many others are trying to hold on to theirs.

What also may have confused some in the media was that just because I was leaving office did not mean I was leaving the Republican Party. In my mind, it is still the party of choice. Is it perfect? No, but no human institution is. And I think you'll see clearly in the pages that follow that I'm as free with my criticisms to those on the right as to those on the left. Any questions or prodding are a direct outgrowth of my hope for this party and my faith in its principles.

But at the heart of all this is something the liberal media has always struggled with: How can a black man be a Republican, a conservative one at that? As you'll see in the pages that follow, it is a question I have had to deal with since I first ran for public office in 1990. To some the circumstance is an anomaly. To me it's just common sense. Still, I am very aware of the perception of me and fellow black conservatives among knee-jerk African-American Democrats and most in the media. It's had repercussions throughout my life, sometimes in ways I never expected.

I can remember quite clearly an incident that occurred in January 2001. It was just days before the new Congress was about to give its official blessing to the election of George W. Bush as president. I had taken the night off and headed south to cheer on my alma mater in the Orange Bowl.

Although it had been more years than I liked to admit, once again the Oklahoma Sooners were fighting for the college football national championship against a first-rate Florida State team. It was a great night for every Sooner fan, and, thank the Lord, it turned out to be another victorious chapter in Oklahoma football history.

For me, it was déjà vu all over again—a return trip to another time, another Orange Bowl, and one of the most important moments of my life—when the Oklahoma team I quarterbacked won the Orange Bowl against another Florida State team back in 1981. But as excited as I was by Oklahoma's win and as great as it was to remember my own Orange Bowl victory, I came back to Washington that night troubled and disturbed by what was a chance meeting.

During the game, I was thrilled to be introduced to Denzel Washington, one of my favorite actors—someone I'd never met but had always wanted to. From what I knew about him, we had some things in common. Both of us were sons of preachers with journalism degrees, we were both family men, and each of us had a special interest in young people—mine as a youth minister and Denzel with the Boys and Girls Clubs.

So, like the true fan that I am, I was excited to meet this man whom I had heard so much about and whose films I had always enjoyed. Despite the noise and the hoopla, we got a chance to visit for a few minutes. We talked about football and his business—movies. I told him how much I liked his film *Remember the Titans,* which I'd seen twice the previous summer. It's a terrific story about the integration of a high school football team, something I knew a little about.

Then we talked some about *my* business—politics. Given Hollywood's animosity toward all things Republican, especially conservative Republicans, I was surprised when Denzel told me that he "kind of liked" George W. Bush. My first thought was that George W. Bush would like him, too, and somebody ought to get them together. "Sounds like just the job for old J. C.," I remember thinking to myself; but just as I was about to volunteer my "matchmaking" services, I found myself hesitating almost instinctively. And then, I realized why.

Despite Denzel Washington's enormous talent and tremendous box office appeal, I held back because I liked and admired him, and I was afraid for him. As a conservative, I knew all too well that the price Denzel Washington would likely pay for meeting with a Republican president would be high—perhaps too high. This world-class actor would risk economic discrimination by Hollywood, personal recriminations by national black leaders, and perhaps even rejection by his fans. His character would be sullied and his blackness questioned.

Once upon a time, I would have introduced these two remarkable people and never given it a second thought. But these days, with the demonization of Republicans in the black community, even a cup of coffee with a conservative president is risky business for an African American in the public eye, so I stepped back. Was I wrong to do so? Maybe, and I have to admit that as I flew back to Washington that night, I was disappointed in myself and frustrated with those who have driven a deep wedge between my people and my politics. I found myself thinking about earlier times when that deep division didn't exist, and I remembered a favorite story about another African American and another Republican president.

On March 4, 1865, President Abraham Lincoln delivered his second inaugural address, unquestionably one of the most extraordinary speeches in our history. There, listening in the audience stood the great orator and freedom fighter, Frederick Douglass, one of my heroes. The president spotted Douglass, a very imposing man, in the crowd, and again, later that day, at a White House inaugural celebration.

Here was one of the most prominent men in the United States at the time, a former slave and fugitive, a fiery spokesman for the abolitionist movement, now a guest at the White House. I can tell you that a visit to the "people's house" is an awe-inspiring occasion for anyone today, so I can imagine what a moment it must have been for Douglass in the waning days of the Civil War when many of his brothers and sisters were still in bondage.

I suppose it's understandable that Douglass was hesitant at first to approach the president, especially after the guards had initially tried to turn him away. But Douglass did join the other guests in the East Room, and later recorded what happened that day.

> *Recognizing me, even before I reached him, he [Lincoln] exclaimed, so that all around could hear him, "Here comes my friend Douglass."*
>
> *Taking me by the hand, he said, "I am glad to see you. I saw you in the crowd today, listening to my inaugural address; how did you like it?"*
>
> *I said, "Mr. Lincoln, I must not detain you with my poor opinion, when there are thousands waiting to shake hands with you."*
>
> *"No, no," he said, "you must stop a little, Douglass; there is no*

man in the country whose opinion I value more than yours. I
want to know what you think of it?"
   I replied, "Mr. Lincoln, that was a sacred effort."
   "I am glad you liked it!" he said; and I passed on, feeling that
any man, however distinguished, might well regard himself
honored by such expressions, from such a man.

No one, Democrat or Republican, would dispute Frederick Douglass on that point. Lincoln stands as a colossus in our history, the original uniter, and the very first Republican president.

Now, fast-forward 137 years and think about this question in today's political environment.

Would Frederick Douglass have accepted an invitation to celebrate the inauguration of the newest Republican president, George W. Bush? I'd like to think so because Douglass was a man of great integrity, intelligence, and compassion.

He was an American, and, whether he agreed with George Bush or not, he would have used the opportunity to bend the president's ear, to argue, as only he could, the merits of his views. Yet, in a kind of ironic reverse discrimination, some African Americans today might actually refuse to cross the threshold of the White House as long as a Republican occupied it, not because of the color of his skin but because of the conservative nature of his politics.

The Congressional Black Caucus is a perfect case in point. After generations of African Americans found the front doors of the White House closed to them, after a 300-year struggle for equality and opportunity, four African American members of Congress actually refused President Bush's invitation to join him, soon after he took office, for a White House discussion of caucus concerns. A tremendous opportunity was sacrificed on the altar of raw racial politics. What an empty and self-defeating protest it was.

Most of us wonder why any American would reject the outstretched hand of a president of the United States regardless of his or her differences. The answer is simple. These four congresspeople, like most African Americans today, are victims not of pervasive racism but of enormous peer pressure to conform to a group identity dictated by a

handful of national leaders who not only refuse to engage in honest dialogue with conservatives but refuse to even acknowledge the legitimacy
of a conservative's right to a different point of view.

I believe too many of these leaders have forgotten their roots and
the conservative values that people throughout American history have
found to be the most effective armor against slavery, racism, and discrimination. Given the past, our progress as a people is nothing short of
miraculous, and it has been values like faith, family, work, and responsibility that have moved us closer to the Promised Land. Yet, African
American leaders shun those who most share these values and, in doing
so, squander opportunities to find common ground that will benefit all
Americans—red, yellow, brown, black, and white.

Like Lincoln, who wanted to "bind up the nation's wounds," George
W. Bush used his inaugural address to bring the country together after
the closest and most controversial presidential election in our history.
And it was to African Americans that he addressed, in the language of a
healer, his most powerful words:

> *While many of our citizens prosper, others doubt the promise,*
> *even the justice, of our own country. The ambitions of some*
> *Americans are limited by failing schools and hidden prejudice and*
> *the circumstances of their birth. And sometimes our differences*
> *run so deep, it seems we share a continent, but not a country.*
>
> *We do not accept this, and we will not allow it. Our unity, our*
> *union, is the serious work of leaders and citizens in every*
> *generation. And this is my solemn pledge: I will work to build a*
> *single nation of justice and opportunity.*

I believe George W. Bush meant what he said. He understands that
the bitterness and suspicion that African Americans feel did not begin
with this election, and more importantly, this division is not the sole
purview of the African American community. African Americans' alien-
ation from and antagonism toward conservatives are only the most visi-
ble symptoms of a larger societal problem facing our country and this
president—the segregation of Americans not by race but by beliefs.

When Lincoln gave his second inaugural address, the nation was nearing the end of a terrible war that drove families apart—a public wound that has healed but still remains tender to the touch. We have come far since then, but the progress we have made as a people is being slowly eroded by a new, more subtle bigotry: a bigotry of ideas, an ideological intolerance that strikes at the very heart of this individualistic society and its values. Today, ideology, not skin color, determines the victims of intolerance and hate, separating us by how we think, not who we are. Just ask Clarence Thomas. Like Judge Bork before him, his fitness for the Supreme Court was challenged because his conservative values collided with the belief system of the liberal elite and the national black leadership.

Like an insidious disease, this new bigotry of ideas is infecting public discourse, social relationships, and political life in this country. It is a dangerous development that should alarm all of us. Equally important, this new intolerance is often cloaked in the basest terms.

Liberals, left-leaning activists, black leadership, and national Democrats have created an unholy alliance where the politics of personal destruction have replaced serious philosophical and political debate. We now live in a world where destroying the careers or ruining the reputations of those we disagree with has become acceptable political behavior.

I've lived through that nightmare personally. Charges of dishonesty, racism, sexism, homophobia, or worse are lobbed with little or no foundation and given credence by a media frightened by the very monster of political correctness it has created. Even "religious profiling," a term coined by Kay Cole James, the director of the federal Office of Personnel Management, has become fashionable from the Senate chamber to the drawing rooms of Georgetown. When it comes to a conservative, no charge, regardless of merit or motive, is too outrageous to leave unreported.

That doesn't mean those on the right are blameless. Over the past two decades, we have seen a hardening of positions, a growing intolerance for opposing views on both the right and the left. I've taken a few hits from a few of my right-leaning "friends," too, and I will talk about that in more detail later; but I believe, contrary to conventional Washington political wisdom, it is on the left that we have seen political rhetoric coarsen and become particularly extreme over the past decade.

Today, liberal black and white leaders level charges of racism with a chilling casualness. Jesse Jackson and Al Sharpton have perfected the art and remain the master teachers whose tactics and twisting of the truth go unchallenged by a media as cowed as corporate America by threats of being labeled racist. Race is recklessly cited as the source of every perceived injustice, from garden-variety incompetence at the Florida polls to honest disagreements on policy issues. Just ask John Ashcroft, whose confirmation hearing was a disgraceful example of my point.

So was the presidential election. The NAACP, a supposedly nonpartisan organization with a proud history, sank to a shameful new low with its now infamous "Byrd" ad in the last campaign. This political smear implied that George W. Bush's failure to support a particular hate crimes bill made him somehow complicit in the horrific killing in Texas of an innocent black man, James Byrd.

Designed to inflame its intended target—black voters—with an incendiary message, it certainly did the job, but at what price? As an African American man, I found this ad particularly offensive. It trivialized a terrible tragedy for base political reasons and created a false view of the world for the sole purpose of frightening blacks into electoral obedience. Ironically, we haven't seen race used as effectively as a political tool since the days of George Wallace when the Alabama governor railed against black Americans to frighten white voters to his racist cause. Both were shameful exercises in the politics of division.

Even after the election, the race baiting continued. Congressman William Clay of Missouri, John Ashcroft's home state, used the Ashcroft nomination for a broadside against the new president. After objecting to the certification of President Bush's election, Clay said, "So much for President-elect Bush's propaganda about reaching out to African Americans," and then went on to actually compare the Ashcroft choice "to the way the Ku Klux Klan members worked to improve race relations. . . . They, too, reached out to blacks with nooses and burning crosses." This kind of harsh rhetoric is completely bewildering to those of us who know George W. Bush, John Ashcroft, *and* Bill Clay.

Admittedly, playing one side against another for political gain is nothing new in American politics, whether it is pitting rich against poor, black against white, young against old, rural against urban, or business against labor. We have seen this divisive political tactic employed from the earliest arguments between our nation's founders to the bombastic rhetoric of Huey Long.

But today, tough but fair political battles and debate have been replaced with a kind of political guerrilla warfare that relies on hit-and-run attacks—leaks, rumors, unsubstantiated charges, and just plain old lies lobbed at opponents in a very personal way. When forced into the open, these self-defined "soldiers of moral superiority" question the motives of their targets and impugn the integrity of their "enemies." Reputations are ruined and lives are destroyed by people who believe that their end justifies winning "by any means necessary."

This politics of personal attack didn't begin with the last administration, but by the time President Clinton and Vice President Gore exited the political stage, they had perfected the practice of ideological intolerance. What were once occasional partisan diversions (for example, the Bork confirmation) had become full-scale personal frontal assaults.

As Americans, I believe, we must reject this new bigotry of ideas and return our nation to one of civility and respect for each other *and* each other's beliefs. When it comes to political views, Congresswoman Maxine Waters and I have about as much in common as a Cornhusker and a Sooner, but I would defend to my last breath her right to her views. Would she accord me the same? It's time we go back to the ideals of Abraham Lincoln; to the dreams of Dr. Martin Luther King, Jr., and to the bedrock values of average Americans like my parents, Buddy and Helen Watts.

I believe the path to this new civility can be found in our roots. The values with which I was raised—values like respect and decency, self-reliance, a belief in God, the worth of hard work, and the strength of family and community—give us a tried-and-true road map to a more tolerant nation, to a better place in our still-flawed yet much more open society; a path to a country in which all of us are judged, to build on Dr. King's words, not by the color of our skin, *nor by the ideology of our convictions,* but by the content of our character.

Labels divide us into warring camps. I'd rather talk in terms of shared values and life experiences than ideology because we're more likely to find unity than division in that way. I am hopeful these pages will paint a picture of where my values came from, and in so doing, show us how much alike we Americans really are, no matter what party we've joined or what family we were born into.

If my American story helps us leave behind the harsh labels so easily bestowed these days—labels like "racist," "Uncle Tom," or "extremist"—maybe we can end up with a higher-level, more polite and humane discussion of the challenges facing all of us today. And in so doing, we can come to understand that different isn't evil; that we can disagree on the solution to a problem without resorting to name-calling.

My values are the result of lessons learned from a loving, caring family and a terrific hometown that presented me with an unusually rich supply of exceptional people who were willing to give a lot of themselves to provide me a firm anchoring, and the strength to pursue better judgment, better actions, and higher goals.

My views on everything from welfare reform to a balanced budget to affirmative action can be traced to what Buddy and Helen Watts taught me as a young boy growing up poor but proud in Eufaula, Oklahoma. My belief in the power of education, the potential of community renewal, and the promise of faith-based institutions comes from an upbringing that taught me the value of school, hard work, and Christ in my life. To those for whom the word "conservative" is a pejorative label, who find conservatism incompatible with the color black, I hope my story will help you understand the roots of my reasoning. Our values and belief systems have a lot more to do with how we were raised and the life we've lived than whether we are Republicans or Democrats, liberals or conservatives.

After four terms in Congress, I am convinced that if we are going to find solutions to the challenges ahead, we must do so in an environment that respects all viewpoints, backgrounds, and beliefs. Ideological intolerance must be rejected as vigorously as racial intolerance if we are going to make this a better country and ourselves a better people, "with malice toward none and with charity for all."

WHAT COLOR IS A CONSERVATIVE?

1

Family Is the Rock on Which We Build Our Lives

Know whence you came. If you know whence you
came, there is really no limit to where you can go.
James Baldwin,
The Fire Next Time, 1962

Donna Brazile, Al Gore's 2000 presidential campaign manager, and I ought to be friends, but we're not. We're about the same age. We're both proud African Americans who are involved in public service and fascinated by the political arena. Both of us can lay claim to a few "firsts"—she is the first black woman presidential campaign manager, and I am the first black to serve in the House Republican leadership.

Our upbringings are also surprisingly similar. Our fathers both did odd jobs, and we were both raised on the poor side of the tracks. Growing up, love was a lot easier to come by than money. We both felt the sting of racism and exclusion, and I suspect that she too was called "nigger" a few times, as I was. Despite a less than easy road, however, each of us managed to graduate from college and go on to make our mark in politics and government.

Yet with all that we share, Donna Brazile decided to use me to play the race card in the 2000 presidential campaign. Attacking my integrity, my motives, and my commitment to African American children, she told a reporter, "The Republicans bring out Colin Powell and J. C. Watts because they have no program, no policy. They play that game because they have no other game. They have no love and no joy. They'd rather take pictures with black children than feed them."

I may have given up boxing years ago, but telling a youth minister he doesn't care about his kids is fightin' words. No one likes to be on the receiving end of a partisan barrage, but if I was going to get shot at, I was glad Colin Powell was in the foxhole with me.

I've never met Donna Brazile. She certainly doesn't know me or what motivates me. Maybe our paths will cross one day, and I'll get to ask her why she chose to single out General Powell and me for what I've come to affectionately call the "full Donna." But knowing her penchant for headline-making remarks (it was an off-hand comment that got her fired as Michael Dukakis's political director in 1988), I found her intemperate comments hardly a surprise.

Ms. Brazile is typical of many black leaders and political operatives today who simply can't accept the notion that there are independent-minded African American men and women who disagree with them, who have rejected their liberal philosophy and approach to problem solving. For them, group identity is more important than individual principles because for them, maintaining the group identity assures the continuation of their power.

Like any group that has endured much, African Americans have created a strong and mutually reinforcing sense of group identity. That's not a bad thing in and of itself. Once the exhilaration of emancipation had worn off after the Civil War and the reality of being free but black in America had sunk in, that group identity comforted, protected, and encouraged us as a people to move forward in spite of the barriers put before our progress. That group identity gave us new strength and courage during the civil rights movement of the 1960s that ended the reign of Jim Crow and began a new era of black educational and economic upward mobility.

Sometimes, though, this group identity can limit us more than it protects us. Just as the Irish, the Italians, and the Jews still take pride in their roots, African Americans today should be proud of their heritage and never forget the difficult path our people have been forced to walk. But when the group identity becomes more important than the individual, it can blind us to valuable viewpoints, options, and opportunities. I embrace my blackness, just as I do my conservatism and my Christianity, but I don't want to be defined or pigeonholed by any one of the many elements that make up my character.

That refusal to be stereotyped and cast into certain beliefs and behaviors is what gets people of color who take another path, particularly a conservative path, into a heap of trouble. It doesn't matter whether it is Colin Powell or Condoleezza Rice, Shelby Steele, Thomas Sowell, Clarence Thomas, or yours truly—we have all been labeled expedients, Uncle Toms, oreos, sell-outs, traitors to our race, and other equally uncomplimentary characterizations.

Most of all, however, critics of black conservatives say we've forgotten where we came from. I may forget a federal budget number or, God forbid, to set the alarm clock for my weekly 6 A.M. flight to Washington, but I know *exactly* where I came from. I know because every decision I make every day is based on the values and lessons I learned growing up on the poor side of the tracks in a dusty little Oklahoma town that most people have never heard of and nobody can spell right the first time.

I like to call the ethos I grew up with "Oklahoma values." But you'd be just as accurate if you said "American values." Except for our lack of a seacoast, Oklahoma has a little bit of just about everything that's American. We call the southeastern corner "Little Dixie" because it touches Arkansas and almost reaches Louisiana. The northern part of the state has a shared border with Missouri and Kansas. The northwestern section is a gateway to Colorado and New Mexico. And, of course, our entire southern boundary defines the northern edge of Texas. We're southern, and we're western, and Great Plains, too.

Every weekend when I fly home from Washington, I never get tired of seeing the beauty and bounty of my home state. I can see the network of rivers—the mighty Missouri and the Arkansas and the North Canadian crisscrossing the Oklahoma landscape. I'm reminded again and again how blessed we are with rolling hills and flatlands of rich, black soil—the kind my family farmed. I never forget, however, that we've also been cursed with drought-ravaged prairies that became the Dust Bowl of the 1930s and created an exodus of Oklahomans that took us 40 years to recover from.

While mother nature may have created the drought, it was man's greed and the federal government that pushed tens of thousands of Oklahoma farm families into poverty. In the depths of the Depression, as farmers suffered terribly from low prices, the government desperately searched for a way to raise farm income. It was trying to do the right thing; but as we see so often in Washington, it came up with the wrong idea at the wrong time. It cut production by paying landowners not to grow cotton, the most common but ecologically destructive crop grown at the time in Oklahoma. It was "King Cotton" that had sapped the life out of much of Oklahoma's grasslands, making them vulnerable to drought and the winds that went along with it. What the feds forgot was that in Oklahoma at the time, it was sharecroppers, not owners, doing the hard labor on most of the land.

When the federal government offered payouts for idling farmland, the owners were happy to oblige, leaving thousands of people financially stranded with no way to feed and clothe their families. Oklahomans learned a hard lesson about the consequences of government "help."

But in those tough times, more Oklahomans stayed than left and the blacks, whites, and Native Americans that make up the majority of our state today remain Oklahoma's greatest strength. The diversity of our people has given us strength of character and sound values forged in the crucible of adversity. It has also given us a diversity of opportunity.

Today in the Sooner state, we drill for oil and gas, we farm and ranch, we mine and manufacture everything from drill bits to peanut butter—and we produce great football teams, too. We have a colorful past and a future of real promise.

So does my hometown of Eufaula, Oklahoma. To find it on a map, look eastward from Oklahoma City and south-by-southeast from Tulsa. You'll see the wavy shape of a lake with some 600 miles of coastline. That's Lake Eufaula, which came into being when I was a kid. The dam that created the lake tamed some very unpredictable rivers and added much more tourism to the local economy, but it also swallowed a lot of rich corn- and cotton-growing bottomland.

Some folks will tell you the lake also swallowed some nineteen burro loads of Spanish gold and silver. Humans have lived around this place where the North Canadian and South Canadian Rivers meet up with Deep Fork River since 12,000 B.C. There were nomadic tribes, possibly following the buffalo that ranged along the muddy riverbanks. In the sixteenth century, Spanish and French explorers and traders passed through as well. The explorers who blazed the road to Texas followed, and later miners headed for California's Gold Rush.

In recent times, treasure seekers have found enough ingots and crosses around Eufaula to convince historians and every kid in Eufaula that the Lost Standing Rock Mine, described on an old Spanish map found years ago in Texas, now lies somewhere under our manmade lake.

The battle of Honey Springs, fought just a few miles to the north of Eufaula, was the biggest Civil War battle to take place on Indian territory. It was also the very first battle in which Native Americans, blacks, and whites fought alongside one another.

Our town's name is from the Creek language, carried west by the survivors of the Trail of Tears in the 1830s. The story of the Creeks, Choctaws, Cherokees, Chickasaws, and Seminoles who made up the five Civilized Tribes, as whites called them, is an indelible stain on our nation's past and a part of Oklahoma's history that teaches us the truth of Dr. King's words: "There are times when a man-made law is out of harmony with the moral law of the universe."

As proud as I've been to be a member of the U.S. Congress, I also know there have been times in our history when our nation has not lived up to its principles. The passage of the Indian Removal Act in 1830, which allowed the removal by force of all Native Americans from the American Southeast to the Oklahoma Territory, was one of those times. Few people raised their voices against this terrible act, passed only to accommodate demands for more land for white settlers and to open Cherokee lands in Georgia to gold mining.

Thousands of Native Americans were driven from their homes, herded like cattle into temporary camps, and then forced to march a thousand miles with few possessions and minimal food and supplies. Four thousand died along the way on this horrible journey—mostly the old and the very young. Yet, as difficult and unfair as their exodus was, thousands more did make it to Oklahoma to begin new lives in a new land. Many rebuilt their nations in and around the Eufaula area. The first Creek settlement in the area became the town of North Fork, which today can be found at the bottom of Lake Eufaula.

I don't know for sure, but I suspect some of my ancestors were among these first settlers. So, when I've risen to cast a vote in the U.S. House of Representatives, my roots remind me that what we do in Congress has the potential to change the lives and the futures of millions of people for better or for worse. That's a lesson I never forget.

North Fork burned in 1863 after Confederate forces were defeated in the battle of Honey Springs nearby. After the war, people began to return to the area, and by 1872, the MK&T Railroad, "the Katy," had set up a terminal to store goods and equipment as a base camp for building a railroad bridge across the South Canadian River to open the way to Texas. A railroad engineer by the name of Bob Stevens gave the thriving little community growing up around the station the name of Eufaula, which is a contraction in the Creek language for "at this place they split up." Two years later, in 1874, another devastating fire repeated the earlier destruction; so, as Americans always do, folks rebuilt one more time.

Like any growing frontier town, once the construction workers arrived, the prostitutes, moonshiners, gamblers, and gangsters weren't far behind, all hoping to make some money off the building boom.

In April 1872, Jacob Cox, the secretary of the interior, arrived in a private railroad car to attend the opening ceremonies for the newly completed bridge. Guns went off all night long. A passerby was robbed of $80 in gold, right next to Secretary Cox's car. Another man was murdered in the camp. That didn't stop the secretary, who spoke to the crowd the next day from the rear platform of his railroad car. As he expounded on Eufaula's great engineering feat, someone fired off a wild shot that whistled past his head, inspiring him to dive back inside. I always think of this as the moment Eufaula began its long tradition of producing outstanding athletes.

The Tenth Cavalry showed up a few days later to restore order to

Eufaula, but the hilly, hickory-wooded countryside was much too accommodating and soon became the hideout of choice for some of the era's most notorious desperadoes. Before long, Belle Starr was a regular in Eufaula, buying supplies for her place up at Younger's Bend on the Canadian River.

Pretty Boy Floyd robbed banks throughout the region in his heyday, and he stopped more than once in Eufaula. On one particular night, he was the only white person at a jumping dance party on the east side of the tracks, not far from where I grew up. Under the light of coal-oil lamps strung on wires from rafter to rafter, a couple of local men began cursing and shoving at each other. Their scuffle was about to erupt into a free-for-all, but Pretty Boy stopped everybody in their tracks by waving a pistol in the air and declaring, "I'm the only bad man here."

My father loved stories about Pretty Boy. There was the time when he begged a meal at a farmhouse and then hid a hundred-dollar bill under his plate. Another time he gave a local black woman $20 to fry him a chicken. For years, there was a small chunk of Pretty Boy's tombstone in my father's bedroom drawer.

Like most of the West, as the train robbers went the way of the buffalo, Eufaula eventually settled down into typical small-town life. When I was a kid, operators were still plugging in local calls one at a time on a big switchboard, something my digital-generation kids can't even imagine. We're all the way up to seven digits now like the rest of the country, but there's still a friendliness to the town that suggests people still talk to the same switchboard operator on a regular basis.

Today, Eufaula is the kind of town where you don't have to be a stranger for long, if you don't want to be. It's a place where most everybody's granddaddy knew everybody else's, and a secret is as hard to keep to yourself as a good fishin' hole. Kids are still pretty safe on both sides of the tracks that run down the center of Eufaula—although not as safe as when I was a young boy racing around town. Those tracks separated black from white in Eufaula's early days, and in some ways, they still do today. But times change. The segregated elementary school I attended, Booker T. Washington, stands broken and empty in an overgrown field. Beautiful red wildflowers grow where small black children used to.

I've always imagined that it was a field of those same wildflowers that greeted a slightly mismatched pair of homesteaders—Charlie and

Mittie Watts, my grandparents, who finally settled on a patch of bottomland in Eufaula and began the Oklahoma branch of the Watts family tree. You would have been hard pressed to find a stronger or more loving matriarch for any family than my grandmother, Mittie. Her straight, gray hair—from a Choctaw relative—gave away her mixed-race heritage that was typical of many early black and white Oklahomans. She had come long ago from Arkansas to live in Talihina, in the Kiamichi Mountains, perhaps 60 miles southeast of Eufaula, on the western fringe of Oauchita National Forest. That's where she met my grandfather, Charlie, an adventurer, who had already buried one wife and was twenty-five years older than Mittie when they married.

Charlie was born around 1865 and also raised in Arkansas, so it's a fairly safe bet that his own parents had been slaves. Some in the family swear that Charlie Watts had a white skeleton or two in his closet as well, but nobody really knows for sure. The records are long gone if they ever existed, and Charlie himself died in 1947, long before either I or my brothers and sisters were around to ask him questions about our roots.

As with so many black Americans, our family history is spotty where it exists at all. There is so much about the Wattses and my mother's family, the Pierces, I'd like to know and probably never will. With children of my own now, I have come to treasure our family stories, seeing them for what they really are—ties that help bind us together in our increasingly impersonal world.

One story has it that in 1912 or thereabouts, my grandfather took off for Edmonton, Alberta, where he did some sharecropping. Why he picked Canada is something else we don't know, but it's likely that someone he knew from home had given it a try and reported back that there were decent opportunities to be had there. Though most black Americans were either former slaves or the children of former slaves at that time, a fair opportunity to work hard and succeed was all they asked for.

But opportunities were limited. In those times, sharecropping was a way of life for many people, both white and black. A poor person farmed someone else's land in exchange for a portion of the crop, plus shelter and provisions. Essentially, it was a carryover from feudal Europe. The landed gentry let peasants work their estates in exchange for a subsistence return, with only the slimmest chance to actually prosper. It also encouraged poor people to raise large families so they would always have lots of hands to work the land. It wasn't unusual, though, for families to work a whole year and end up with nothing, or even in debt, depending on how farm market prices turned out or how honest their landlord was.

Charlie Watts had a brother living in Talihina, and eventually his brother convinced him to come down and invest his earnings in land there. He bought a farm, met and married my grandmother, and the two of them worked that land for perhaps three years. They sold it sometime around 1916 or 1917 and headed back to Edmonton with their young son and daughter. Another son, Earl, was born in Canada, along with another daughter. Still another child, a daughter, died in infancy.

Five more years of Canadian sharecropping followed until a bad year up north convinced my grandparents, now with a growing family to feed, to look for better prospects. In 1922, they settled in Eufaula, where they bought a 15-acre bottomland parcel. A year later, Mittie gave birth to a boy, and they named him J. C. The initials didn't really stand for anything. Many Southerners are given names made up only of initials, but before long everyone called him "Buddy." Ironically, if it weren't for a bad crop year, I could have ended up playing football in Canada as a Canadian.

My daddy was a firecracker from the day he was born to the day he died, and not a day goes by that I don't miss him. I'd love to have seen him when he was a young boy. From what I've been told, Buddy was the real reason for Mittie's gray hair.

One of my favorite family stories, for obvious reasons, is the one that explains his name. It seems that when Buddy was in third or fourth grade, his teacher asked him what "J. C." stood for. "It doesn't stand for anything," he told her.

"But it has to stand for *something*," the teacher insisted.

And in an instant that would later determine my name, and that of my son, J. C. Watts III, better known to us as "Trey," Buddy remembered that one of his sisters was reading about the Roman Empire in school. So he proudly said, "Julius Caesar!"

Buddy was next to youngest in a family of nine kids. Back in those days, big families were quite common. With a workable piece of bottom-land, Charlie and Mittie did pretty well for a black family in that place and time. Along with farming, they also built up a herd of about 23 dairy cows. They needed a lot of hands just to keep that many cows milked.

Of course, the Depression and the infamous Dust Bowl years weren't far away, and they were pretty tough times for my family just as they were for most everyone who stayed on in Oklahoma. Daddy used to tell us that Mittie worked for $10 a month, while he and his brothers worked ten long, hard hours a day to earn 75 cents. He also remembered that Mittie knew how to save enough from her wages to give all of her kids a nice Christmas.

When the price of apples fell dramatically during the Depression, a local farmers' cooperative let their harvest fall to the ground, then dumped truckloads on a trash pile not far from Grandma Mittie's. My father and his brother went down with tote sacks and picked out the best of the best to bring home, wondering all the while why the farmers wouldn't just let hungry local people come over to pick their apples.

You don't forget hearing stories like this growing up. Years later, my older brothers, Lawrence and Melvin, would pick through that same trash pile for salable pieces of scrap iron, salvageable used tires, and glass bottles that would bring them a few pennies apiece. The Watts family was into recycling long before recycling was cool.

In what turned out to be the last of the drought years, my grand-father got only $100 for his corn crop—exactly enough to pay for the seed it had taken to grow it. Of course, he needed another $100 for the next year's seed as well. That meant coming up with fresh capital from somewhere. The Watts's dairy herd became the family's "cash cows."

With money in short supply everywhere, the owner of the local gro-cery store was one of the very few people with what we call "liquidity" today. He picked out the ten best cows in Grandfather's herd and paid him $10 a head. It may seem like a bargain now, but it gave the family just enough to go on until better times came around.

Folks during the Depression had a saying: "When you can't make it from the crops in the field, better have a still on the land." My uncle Earl took it to heart. In 1937, Earl was courting a girl and conjuring up a little moonshine on the side when the federal "revenuers" broke down his door and hauled him off to jail. Although he started a prison term for selling bootleg whiskey, the authorities eventually found out he'd been born in Canada and deported him.

While what he did was wrong, his life remains one of the sad chapters in the Watts family history. Because of his conviction, he was never able to come home again.

Another tragedy struck the Watts family when Daddy was about 14 years old and still living at home. His sister, Mattie, developed tuberculosis in the summer after her graduation from high school. She was the first Watts to get a diploma and the pride of the family. Black families have always understood the value of education, and the Wattses were no different.

From the picture I've seen, she was a beautiful girl and the apple of my father's eye. Back then tuberculosis was a life-ending disease, and despite her family's heroic efforts to save her, the illness sapped her energy day by day as her younger brothers and sisters watched.

They did the best they could in a home with no electricity. Coal-oil lamps were the only source of light. There was no air-conditioning, of course, and as her fever rose, her father and brothers carried her outside to lie beneath the shade of a big tree. Her brothers and sisters took turns trying to keep her cool with makeshift fans fashioned from old cardboard boxes.

Despite everything the family did, Mattie got worse. The only option was to move her to the tuberculosis hospital in Clinton, about 80 miles west of Oklahoma City and far from the family that loved her. But my daddy had a contrary streak. He wasn't about to let his sister sit alone in a hospital miles away, so he grabbed a freight train and rode the rails to Clinton to spend the two-hour visiting time with his favorite sister.

With no money to speak of, he spent his nights in Clinton, too, sleeping under a bridge, waiting for visiting hours to come around again. But despite Buddy's love and the family's prayers, Mattie died a few months later.

That story of my daddy's train ride to his sister's bedside taught me the importance of love and family from the time I first heard it as a young boy. No matter how much we change or how far from home we find ourselves, the family is the rock on which we build our lives. My family is a proud one—proud of its ancestry, its triumphs over adversity, its faith, and its love.

I know exactly where I come from. Where I come from, I was taught that I didn't spend out more money than I took in. I was taught that education was important. I had to go to school and act civilized when I got there, respect my teachers and all adults, work hard, and exercise personal responsibility. I was taught that doing the right thing matters a lot. My first and best teachers were Helen and Buddy Watts, with a back-field of grandparents and aunts and uncles and family friends that got me where I am today. I know that they will always be there to help get me wherever I go tomorrow.

2

Opportunity Is Just Hard Work in Disguise

A truly American sentiment recognizes the dignity of
labor and the fact that honor lies in honest toil.

Grover Cleveland,
in a letter accepting the nomination
for president, August 18, 1884

I tell anybody who wonders where my strong belief in the value of hard work and self-reliance comes from that I learned it at my father's knee, and sometimes over it. His life was a testament to what we can accomplish, regardless of the circumstances of our birth, if we look at life as victors, not victims.

Believe me, my daddy understood the barriers that racism and segregation put in his path. He used to say, "The only helping hand you can count on is the one at the end of your sleeve."

He was no Pollyanna. No one who lived through the Depression, World War II, and Jim Crow could be. He would have had plenty of reasons to throw up his hands and blame his troubles on others, but he never did, because Buddy Watts believed—and taught me—that opportunity is just hard work in disguise.

Certainly, he raged at the racism around him and fought it in his own way, but he also refused to let it keep him from succeeding, from taking care of his family and moving up in the world.

Arthur Ashe once said, "Racism is not an excuse not to do the best we can." Buddy Watts was living that truth long before that great tennis player hit his first ace. Daddy was only 13 years old, barely out of short pants, when he and his younger brother, Lois, only 10, moved out of the house and took up residence in a log cabin their father owned across the creek. They still had to work the farm for my grandfather, but when it came to food to eat and clothes to wear, they were on their own at an age when most kids today haven't faced anything more serious than the loss of television privileges. Lois says they sometimes lived on 15 cents' worth of rice a day. While Buddy seemed to thrive on his newfound independence, Lois missed his mama's cooking.

Buddy was a tough customer even as a kid. So tough, in fact, they used to call him "Frostproof" because he liked to go hunting in the snow barefoot. Lois claims that Daddy could strike a match and hold it under his feet until it burned his fingers, and it still wouldn't burn his feet.

I guess Daddy just had a thing about shoes as a kid. Mittie would send him off to Sunday school all properly dressed for church; but as soon as he was around the bend and out of sight, off came the shoes. I can just imagine Buddy marching into church in his Sunday suit with his bare feet flapping on the old wood floors. He was a character.

During the Depression, money was so scarce that my father dropped out of school altogether after only two days in the seventh grade and started working full-time—something that he regretted the rest of his life. But in spite of his lack of formal education, his determination to succeed was so great that he owned his own house at the ripe old age of 14—bought from his father.

Buddy farmed, did odd jobs, worked construction, and even boxed a little. Somehow, he managed to put together enough money to buy a 1937 Terraplane, a sleek coupe with plenty of power under the hood. One night, he and a drinking buddy were out driving south of town. He swerved to miss a bridge abutment that had suddenly "surprised" him. Daddy claimed it was his friend's fault—the friend was supposed to be navigating because Daddy had told him that he was too drunk to see where he was going.

When the bridge rose up to meet the wild pair, the Terraplane keeled over onto one side and skidded to a rough stop. Its windshield popped out and shattered. Luckily, Buddy and his friend only got the wind knocked out of them and were able to walk back to town, figuring the car would probably be impounded. The next day, he learned the police had simply righted his car and left it, now in less than prime condition, on the shoulder of the highway.

Nothing stopped Buddy, however, not even the loss of a windshield. He bought a pair of goggles at the local hardware store and roared off with the fresh air "cooling" the Terraplane as he went. It took a swarm of bees, encountered midflight on the way to Tulsa, to convince him of the value of a windshield.

Like most people at the time, Buddy married quite young, at age 17. He and my mother, Helen Pierce, tied the knot in 1940 and began a life together that would see the birth of six children and last until her death in 1993. Despite my mother's patient influence, in the early years of their marriage, Buddy could still manage to get himself into a heap of trouble—but at least he was in the doghouse for all the right reasons.

In those days, he was making a whopping $3 a week, so he was always looking for ways to add a little income to the family treasury. One moneymaking operation finally landed him in the pokey. Buddy would take one-third of his earnings down to the grocery and buy malt, hops, yeast cakes, and sugar. I'd like to tell you he was baking cookies on the side, but he wasn't. He was making an alcoholic delight known around Eufaula as "cholk beer." A dollar's worth of ingredients and three days of brewing magic, and my daddy had a product that could be sold—illegally, of course—for $1 a gallon.

Just a dollar in cholk beer sales fed the growing Watts family for a week. But the problem—and the risk—was keeping it a secret. When the town drinkers beat a pretty regular path to one door, the local police were rightfully suspicious that they weren't there for a prayer meeting. My daddy later became a preacher, but these were the days before God explained to him the evil of cholk beer.

Daddy would tell his customers to be discreet, but folks who down cholk beer are not usually known for their prudence. One night, a bunch of men were hiding down by the creek bed doing a little drinking when they ran out of beer. They came knocking on my daddy's door demanding more. The problem was, they didn't have the money to pay for it.

They started raising such a ruckus that Daddy had to do something to get them to quiet down. So he picked up a quart bottle and led them back to the creek. He had a plan. He offered to wrestle the strongest one among them. If their man won, they'd get the bottle for free, he told them. The match got underway, and Daddy eventually lost. What the drinkers didn't know was that as he held their "champion" upright, pretending to struggle, he was going through his opponent's pockets for enough change to pay for the beer.

The penalty for making and selling cholk was known as "30/50": thirty days in jail and a $50 fine. Daddy paid the price one time, and he soon learned that the driver who took him and his fellow cholk "entrepreneurs" out to work on the county roads was usually drunker than any of the road crew had been at the time of their arrest.

My parents always understood that the path up and out of poverty meant faithfully putting a little something aside to invest in the future, not expecting government or someone else to do it for you. When it came to scrimping and saving, I'd pit Mama's ability to manage a budget against anyone in the Congressional Budget Office. Somehow, between the cholk beer and a lot of hard work, they scraped together enough extra money to buy another house, which Daddy leased to a man and his wife for $1.25 a month. It had only one room. Buddy added a second, which doubled the size of the house, and he raised the rent up to $2.50. The Watts real estate empire—Daddy's dream—was born.

The couple went along with the increase, but like so many of Daddy's tenants over the years, they needed a couple of extra weeks some months to get the rent money together. That's how tenuous and tough life on the poor side of Eufaula was then and, often, still is today.

Eufaula could hardly have been called the land of opportunity in those days. It wasn't big enough to offer many high-paying jobs, so Daddy was always on the lookout for opportunities and always willing to travel. In 1951, he and my mother took off for Chandler, Arizona. Rumor had it that local growers needed experienced cotton pickers, so my parents hit the road in search of honest work.

And it was in Arizona that Buddy finally met his match. He hadn't picked his second sack of cotton when a rattlesnake up and bit him. He spent some touch-and-go days in the hospital, but even a rattlesnake couldn't put Buddy Watts out of action long. He was just too tough to die.

As soon as Daddy was back on his feet, he and Mama were gone again, this time to Bakersfield, California, where a recent earthquake meant there were plenty of construction jobs for those willing to work. That's how it was for my family in those years: following the jobs and moving from place to place, which was hard on a bunch of young kids. Sometimes in the winter, we stayed a little longer in places like Milwaukee and Wichita, but spring and summer always meant going home to Eufaula—back to farming the land, back to fresh air and good food.

Then, the harvest would be over, and it was back to a big city for the Watts clan, traveling like a band of gypsies, looking for the best-paying construction jobs. Buddy and Helen always rode up front in the cab of the old GMC stakebed farm truck. We kids rode in the back, holding down the furniture and sometimes holding on for dear life.

As second youngest child, I was luckier than my older brothers and sisters who spent years on the move. For me, the nomad existence ended not long after first grade. In the 1960s, with so many children to keep track of and keep in school, Mama began to stay in Eufaula through the winter with all of us while Daddy went off to find work.

We missed him during those cold months, but Mama kept us warm and fed and gave us enough love to cover Daddy's long absences. It was a difficult life for my parents, but eventually, through all this job chasing and after some determined saving, my folks had more than twenty rental houses around the neighborhood. When I was a kid, the number usually hovered around four or five. Daddy always kept an eye out for nearby low-cost properties coming on the market. Most of them gave new meaning to the word "fixer-upper." Daddy's years of construction work finally paid off as he used his experience to renovate more and more old houses; but now he was working for himself.

Buddy Watts understood the concept of diversification even if he didn't know the proper word for it. Along with construction work, Buddy put together down payment money through other sideline businesses like hauling hay and watermelons and furniture. As we got older, my brothers and I supplied a lot of the muscle, but Daddy never asked us to do anything he wouldn't do himself.

At one point in his life, he even tried his hand at the food business. On a spot overlooking the lake where Mittie's house had been, Buddy built a concrete-block barbecue stand. It's still there today, run by my brother Lawrence. Just drive over to Lake Eufaula and follow your nose in the direction of the rich hickory smoke. I may be a little biased, but I think Lawrence serves up the best barbecue in the state.

Buddy didn't stop with his barbecue, however. He also built a place on our property for an after-school teen hangout, complete with a jukebox. I helped pay his bills during my high school years by feeding it one quarter after another to hear Marvin Gaye sing "Let's Get It On" over and over again. I knew every word and fancied myself every bit as good a singer as Marvin. Thinking back, I'm just lucky Daddy didn't charge me extra for the noise he had to put up with.

Having your father own the local teen hangout was great. But having your dad as one of the town cops definitely had its drawbacks. In 1969, at age 46, my father became Eufaula's first black policeman. For Buddy, it was another opportunity he couldn't turn down—although he nearly did.

The city fathers wanted him to limit his patrols to the eastern side of the tracks, the black community. I give them credit for being forward-thinking enough to want a black man on the force. This was only a year after the murder of Dr. King, and racism was still alive and well in Eufaula as it was in much of the country. Unfortunately, the town leaders were also afraid of the repercussions if a black policeman ever tried to arrest anyone white.

By middle age, Buddy may have sowed all the wild oats he was going to, but that didn't mean he was any less contrary. He flat out refused.

"No," he told them. "I'll work for the police force, and I can be a good cop. But I'm going to be a policeman for *all* of Eufaula. I want to be able to arrest anybody who's breaking the law."

And he did. It was speeders, black and white, who really got under Buddy's skin. Anyone willing to risk the lives of children to save themselves a few minutes could expect no mercy from Buddy Watts. He could pull out a ticket book faster than Pretty Boy Floyd could unholster a gun. One of his favorite jokes was about a cop who pulls a guy over for slowly cruising right through a stop sign. The driver gets out of his car and demands to know why.

"You didn't stop back there," the cop says.

"Yeah, but I slowed down," the driver answers. At that, the cop starts whopping him on the head with his club.

"Ow! Stop!" the driver pleads.

"Do you want me to stop," the cop asks, "or should I just slow down?"

My daddy took on a lot of different jobs in his day, but I think he really loved being a policeman. There were a few people in both the white and black communities who didn't particularly like Daddy, but almost everyone respected him. One time he stopped a black man for making the same mistake as the guy in the joke—running a stop sign.

The man took exception. He went out and bought himself a handgun to keep in his glove box. "The next time Buddy Watts stops me for some bullshit," he told people, "I'm going to show him."

Before he could make good on his promise, he got into a fight with another man, and when he drew the gun supposedly meant for my father, he was shot dead—not 50 yards from our house. The man who did the shooting came right over and turned himself in to Daddy.

Buddy served 10 years on the police force until a heart attack sent him into the hospital for a double bypass. That was a tough time on the family, but I think it was harder on Buddy than anyone else. My daddy was a bear of a man who had always defined himself by his physical strength and his ability to provide for his family. I saw Daddy pick up refrigerators by himself, and my coach once told me he saw Buddy carry a freezer on his back.

All of a sudden, there he was, weak and sick, lying in a hospital bed with tubes in his nose and his chest split from chin to waist. For the first time in his life, my father was totally helpless, and he hated it. I imagine the nurses that pulled "Buddy Watts duty" could probably apply for sainthood today. Patience wasn't one of his virtues.

When the doctor finally gave him his walking papers, he headed home. I stood in the yard that day waiting for him to arrive, ready to give him an arm if he needed it, but I really didn't think he would. The father I knew would walk proudly on his own back into his house. But he surprised me that day. As he got out of the car, he covered his face with his hands and began to cry.

"I didn't think I was going to see home again," he said. It's a disquieting moment for any child to see a parent as human, with human weaknesses. It was good for me and probably good for Buddy, but it didn't last long.

He spent most of his recovery worrying about his rental houses, and he was back on all his jobs long before he should have been. But that was Buddy.

Ten years later, he faced a second heart surgery. Remembering the pain of the first, he asked the doctor, "Hey, Doc, how many patients do you lose in this kind of surgery?"

"About 1 in 200," the doctor told him.

Buddy sat there for a moment, thinking, and then he said, "What number are you on, Doc?"

This good-natured ribbing continued on after the surgery. As Daddy was recovering for the second time, the doctor told him he was going to have to change his eating habits.

"You can't eat the way you have been, Buddy," the doctor lectured him and started to go down a long list of future no-no's.

But it was when the doctor ruled out eggs that Buddy sat up and took notice.

"Can't eat eggs?" he asked.

"No, you can't," the doctor told him.

Daddy wasn't giving up. "How many can I eat in a week?"

"Well, you shouldn't eat more than a couple a week."

Daddy sighed. "By golly, Doc, there was one morning I ate thirteen."

Most of us in the family believe Daddy probably worked himself into that hospital bed. Taking a job on the Eufaula police force didn't mean he cut back on his other work—he just slept less. While serving on the force, he would start his police rounds about 10 at night, get off at 7 the next morning, then sleep for two or three hours before going to work on his rental properties, or hauling hay, or helping folks move. He'd get back home around 4:30 to take a bath, enjoy his supper, watch the evening news (a must), and go to bed until it was time to hit the beat again.

But as if being a handyman, a real estate "baron," a barbecue entrepreneur, a hauler, and a policeman weren't enough, Buddy Watts was also a man of the cloth. He may have strayed from the straight and narrow as a boy, but he spent most of his adult life trying to keep the rest of us on the right path.

I've never really understood how or when my father's conversion from a good-natured hell-raiser to a God-fearing preacher came about, but faith became a central part of my father's life, and ours as well.

On Sundays, he'd pastor a small-town church—and sometimes more than one—somewhere within a 60- or 70-mile radius of Eufaula—little towns like Vian and Spiro. First and third Sundays, you'd find Buddy preaching at one and the second and fourth Sundays, he preached at the other. That didn't mean we'd all pray together. Most often the family would head to the Sulphur Springs Baptist Church with Mama at the head of the Watts parade while Daddy headed out into the countryside to preach to the faithful. In country churches, it's not unusual for a circuit preacher to do the pastoring duties.

Daddy used to tell us about one church he preached to in a town called Canadian that had just six people. But when it came to sharing the Lord's message, size wasn't an issue with Buddy. I guess the congregation appreciated his commitment because they each donated $2 to their visiting pastor, which was a real sacrifice for this mostly poor black congregation.

Faith in Christ was central to my father's existence, and seeing the depth of his faith helped me find my own. I can't say we were close when I was a child. Daddy didn't have time for intimate relationships with his children, but as he and I grew older, we began to talk about all the things that were important to us as men, as fathers and husbands, and as providers. We also talked politics, but we never argued.

Daddy may have been a dyed-in-the-wool Democrat, and me a rock-solid Republican, but we were both conservatives and shared the same values. I always laugh when I remember one of Daddy's best-kept secrets—this lifelong Democrat activist stayed in the political fold locally but voted for Republican presidential candidates for years.

In the end, I know he was proud of me and what I have accomplished—on the football field and in politics. But it was my relationship with Christ, the importance of faith in my life, that let us find real peace with each other and, in doing so, build a deep bond I still feel today.

Daddy died October 26, 2000, not of a heart attack but after a two-month bout with cancer. As you might expect, Buddy never let up even when fighting a life-threatening disease. About a year before he died, I was home in Eufaula and wandered over to his house. No visit home was complete without spending a few minutes "taking Buddy's temperature" on the issues of the day, and Buddy always had something to say about Washington, whether I wanted to hear it or not.

I poked my head inside the front door, but he wasn't around. I knew he was home somewhere, so I headed for the back shed, and that's where I found him; 76 years old, dressed in a flannel shirt and denim overalls; at the top of a 12-foot ladder pounding nails in the ceiling. Daddy had decided to winterize the shed for his latest pride and joy, a huge RV he liked to drive across the country from time to time.

I stood there for a minute watching him scoot his big frame up the ladder, pound a few nails in, and then try to "walk" to the next point on the wall by brute force without getting off the ladder, of course. It was a strange moment to find myself having to lecture my daddy, of all people, on ladder safety.

I don't think he paid me much mind, but then this was the same man who was still hiring out to clear stumps and harvest crops for local farmers with his big, red Massey-Ferguson tractor. When Buddy was riding that big machine, he was on top of the world, and he wouldn't have been any happier if he were behind the wheel of a Ferrari. If I'm half that active when I'm his age, I hope the Lord will give me the good sense to retire in peace.

I finally got Daddy to come down off his ladder and sit a spell with me. But before we could put our feet up, one of his best customers pulled into the driveway in an old pickup, coming to settle a bill for the brush Buddy had cleared for him the week before.

Before the man's boots touched the ground, Daddy gave me a sly wink and said, with a twinkle in his eye, "Here comes money!"

3

No Matter How Far You Go in Life, Someone Else Helped You Get There

A mother is not a person to lean on but a person to make leaning unnecessary.

Dorothy Canfield Fisher

When championship runner Wilma Rudolph was 4 years old, she contracted double pneumonia, scarlet fever, and polio, which left her unable to walk normally. For years, she wore a heavy leg brace, but it was her large and loving family that gave her the moral support and the daily physical therapy she needed to beat her disability. After nearly six years of painful rehabilitation and even more determination, Wilma Rudolph threw away her brace one Sunday morning and, to the amazement of the congregation, proudly walked down the aisle of her family church in Clarksville, Tennessee.

In the 1960 Olympics, she became the first American woman to win three gold medals in track and field and the title "fastest woman on Earth." As an athlete, I've always loved Wilma Rudolph's story of individual triumph over tragedy. Her nearly impossible challenge would probably have defeated most of us, but this courageous young woman had a secret weapon: a loving, caring mother who helped her daughter believe in herself and in her ability to reach for the stars.

"The doctors told me I would never walk again," Wilma Rudolph said in explaining her near-miraculous recovery, "but my mother told me I would, so I believed my mother."

Like Wilma Rudolph, I was lucky to have a mother who believed in me, too, and in my dreams even when they took me down a different path. Just as my father did, Helen Pierce Watts gave me the values and the faith that sustain me today.

"The roots of responsibility and the wings of independence" are what motivational speaker Denis Waitley calls a parent's "greatest gifts." My parents gave me both, but each in their own way.

If Buddy Watts was like a strong wind in our lives, a whirling dervish of a man, pushing and pulling the family forward with his ambition and can-do attitude, my mother was the sheltering oak, steadying us in the wind and giving my brothers and sisters and me the stability and love all children need to grow.

Daddy made the living, but it was Mama who determined the way we lived. With Buddy gone on one job or another a good deal of the time, Mama was the glue that kept us together. I'm still not sure to this day, however, what brought the two of them together.

With the country still recovering from the Great Depression and the world at war, Helen Pierce married Buddy Watts in a civil ceremony at the McAlester, Oklahoma Courthouse. Though he'd been on his own for several years, truth be told, Daddy was actually a little younger than Mama when they "jumped the broom," and he didn't have much more to offer her than the sweat of his brow. Given his reputation at the time, I imagine ol' Buddy probably swept her off her feet.

When looking for a wife, Buddy didn't have to go far. Mama grew up in Eufaula and came from a big family like Daddy's, with six brothers and sisters. Once married, it didn't take her long to produce a family of her own. By the time I came along in November 1957, Mama was supervising my older brothers, Melvin and Lawrence, and my older sisters, Mildred and Gwen. Grandma Mittie was still down on her cotton-growing land a few blocks away, but she was always there to lend an extra pair of hands when they were needed.

When Mama's last child and my baby sister, Darlene, arrived a few years later, the Watts family was complete, and Mama served as presiding officer. Like Grandma Mittie, whom she occasionally butted heads with, Mama was an independent woman of iron will, both mentally and physically strong even though she suffered from a bad limp, the result of a car accident. I don't think the idea of giving up on anybody or anything ever occurred to her in her life. She certainly never gave up on any of us in good times and bad.

One of my mother's great strengths was her ability to balance our desire for independence with the discipline that every child needs. Mama kept all of us on a pretty short leash, and now that I'm a parent of teenagers, I understand why. Left to our own devices, the Watts kids could stir up plenty of trouble even in a little town like Eufaula, but with Mama in charge, we knew there was always a price to pay for straying too far from a righteous path.

She was tougher than most of the linebackers I faced, and when Mama put her foot down, every one of us knew she meant business. Growing up, I don't think a month went by that somebody didn't wear out a pair of shoes, but the six of us together never wore down this rock of a woman who, despite our occasional missteps, was always there for us.

Helen Watts was never a demonstrative woman with any of us. We weren't showered with hugs and kisses as kids, but no one loved her children more than my mother. Henry Ward Beecher, the great abolitionist preacher, said, "The mother's heart is the child's schoolroom." He could have written those words about my mother. She had only a tenth-grade education, but she was wise in the ways that matter—teaching us right from wrong, giving us confidence in ourselves, and protecting us from harm. But perhaps most of all, she was an example to us.

I'm sure there were moments, maybe a lot of them, when the weight of keeping the Watts family going must have been tremendously difficult, when she was bone-tired or worried about money or one of the kids. Yet, she shouldered her burdens without complaint and lived a life of sacrifice, self-reliance, and faith in herself and God that became a lesson learned for me. Whatever civility I have, it is because my mother taught me well with the example of her own life.

Looking back now, I realize what a hard life that was: Moving from house to house as Daddy bought more rental properties. Raising six kids with a husband gone as much as he was home. Trying to make ends meet and feed a big family without much money.

Besides keeping myself and my five brothers and sisters out of trouble, her biggest challenge was just keeping the Watts family afloat. Despite Buddy's nearly round-the-clock work habits, money was always tight, and Mama's job was to make what money we had go as far as it could. She cooked food that would stretch to fill a houseful and leave something for leftovers. No family in America can claim to have eaten more beans than the Watts clan. Once in a while, there was something special, like pork chops or fried chicken, but just as often there wasn't money to buy enough for everybody.

My mother had a solution. When she put the platter on the table, she often didn't take a helping for herself. Like most kids, I was usually so busy trying to get my share I never realized that Mama rarely got hers. Instead, she would sit quietly on the couch, reading a magazine or doing the mending while we noisily plowed through Mama's good cookin', never giving her a second thought. Afterward, I would see her in the kitchen, cleaning up and making her meal out of what she found left over on our plates.

Years later, as a parent myself, the memory of her sacrifice made me feel ashamed that I had not appreciated her more then. I suppose most children are like that. They assume parents are there to provide for them, and they are. We owe our children because we brought them into the world. My mother understood that special responsibility all of us as parents must shoulder. She made sure we were taken care of emotionally and financially, but it wasn't easy.

Her ability to budget the little money that Daddy gave her for groceries was uncanny. Every time she sent me off to the store with a shopping list and a twenty-dollar bill, my hopes would soar that there'd be enough left over for a Hershey bar and a soda.

As I high-tailed it down the dusty road to Burns' or Knight's Groceries, my mind would be racing twice as fast as my feet. "Would this be the shopping trip when Helen Watts's vaunted accounting system would fail her and the Hershey bar and soda pop would, at long last, be mine?" The answer was always the same. By the time I'd rounded up every item on the list with absolutely no extras, I'd be lucky to walk out the door with 15 or 20 cents in my pocket. I soon learned that going to the store for Mama was not the get-rich-quick scheme I thought it would be. It's no wonder that I've been a "take-no-prisoners" conservative when it comes to balancing the federal budget.

Now, Daddy was another story. When I went to the store with him, I could slip a bag of cookies or a half-gallon of ice cream into the cart. Not so with Mama. Still, like Mittie, she found ways to put aside a little bit here and there from the household budget, and whenever one of us kids needed something special for school, she found the money. When I was in college taking my first journalism class, I desperately needed a typewriter. I knew I couldn't squeeze the $300 price tag from my spending money. Mama came to the rescue, and to this day, I don't know how she did it. But I am still grateful because I wouldn't have dared asked Daddy. His "extra" money always went for another down payment on another rental house.

Helen's rainy-day fund was there for her kids, but she rarely spent money on herself. Even though she had a big family, Mama never had a fancy washing machine, much less a dishwasher or a microwave oven. My kids today couldn't tell the difference between a wringer-washer and Uncle Earl's old still, so they can't appreciate what hand-washing the clothes for a family of 8 really means, but I remember Mama bent over that old washer for hours on end wringing our clothes out by hand. It was backbreaking work.

I've always wondered if that wasn't part of the reason why we rarely got new clothes. The more clothes we had, the more Mama had to wash. Helen Watts was nobody's fool. When I got older, my mother taught me how to wash my own clothes and how to cook—another lesson in independence that comes in handy today.

With money in short supply and health insurance nonexistent for most of the farm families in Eufaula, nobody saw the doctor unless they were in childbirth or at death's door. Whenever any of us kids got sick, Mama had her own "in-house pharmacy" of old home-style remedies ready to go. They didn't have official FDA approval, but the women of Eufaula had thoroughly tested their effectiveness on several generations of guinea pigs, namely, their children. I don't know if any of these cures really worked, but they've been passed down for decades.

When kids were teething, for example, people would drill a hole through a dime and tie it around the baby's neck with a piece of string. The silver in the dime supposedly helped the baby teethe easier. I have my doubts, but I saw a heck of a lot of black babies in Eufaula wearing "FDR lockets" on a fairly regular basis. When it came to toothaches, Mama had a quick if not completely painless solution. More than once, I had my baby teeth pulled out with a string in Mama's gentle hand. In fact, I didn't see a dentist until I was in eighth grade.

For the mumps, Dr. Mom prescribed a cotton diaper soaked in sardine oil and other equally fine-smelling poultices. Mama even had her own home-remedy cough syrup: rock candy and peppermint with lemons and whiskey. We used to laugh that if it didn't cure you, it would kill you!

Having grown up without the security that health insurance gives a family, I have a perspective on health care that most politicians don't. I've lived the problem, and even though my mother and so many others like her on the east side of the tracks in Eufaula were pretty talented at keeping their kids healthy, poverty takes its toll. My aunt who died of tuberculosis is a perfect example of the price the poor must pay for a lack of good health care—a problem that continues for many today.

Whether it was for the doctor or clothes or food for the table, money was always in short supply for the Watts clan, as it was for most black families in Eufaula, but we were luckier than most. My mother never worked outside the home, and growing up, we took comfort in knowing Mama would be there to welcome us home from school when we came racing through the front door.

My views on child care, one of the most serious issues impacting family life in the United States today, come from my own experiences growing up in Eufaula. I believe the presence of my mother in the home made a major difference in my life, but I also saw many mothers in my neighborhood trying, under difficult circumstances, to hold down a job and keep their families together.

Unlike the late 1950s and 1960s when I grew up, women have many more opportunities for careers today. But for the moms in my neighborhood then and now, the decision to work outside the home wasn't made by choice but by necessity. It wasn't easy for any of them, especially for the single mothers who struggled mightily to keep food on the table and their kids in line and still play a loving, nurturing role. When women are left to handle the responsibility of two parents, the lack of a father takes a heavy toll. I saw that growing up too, sometimes in the lives of my friends.

Between taxes and the cost of just about everything these days, more and more parents must juggle work with the needs of their kids. With five children, it's been tough for Frankie and me these past eight years, but especially for Frankie, who has to shoulder more burdens than most moms because I've been in Washington four days out of seven most weeks.

I'm not a fan of federally regulated or controlled child care, but the country as a whole, and many communities, are getting better at developing local private, public, and faith-based child-care programs. Many of these programs are excellent, and while they can never substitute for a parent, they can provide support for children and parents who need it.

I've always thought that our retired seniors, grandmothers and grandfathers who are living longer, healthier lives, could also be a big part of the answer to our child-care problems. Alex Haley got it right when he wrote, "Nobody can do for little children what grandparents can do. Grandparents sort of sprinkle stardust over the lives of children."

No one has more experience with children than grandparents. Why don't we use it? If we don't, we are wasting a precious national resource that could make a huge difference in the lives of millions of children.

I know, because my Grandmother Mittie certainly made a big difference in mine. In fact, after my mother, she was the most important woman in my life growing up. Most adults have tried at one time or another to retrieve their earliest memory of childhood. My first memories are of Grandma Mittie Watts. Eufaula was the kind of small town where the kids could run the streets in perfect safety. It wasn't unusual on a summer day to see a 6- or 7-year-old head out the door to play at a neighbor's or visit a grandmother a few blocks away.

I remember running off to Mittie's house near the shores of Lake Eufaula whenever the spirit moved me, and of course Mama said,

"Okay—git!" Mittie's place was really an extension of our own home. I guess you could say the Watts family had two front doors—they just happened to be a 10-minute walk apart.

Looking back, it seems to me that I lived at my grandmother's place as much as I did my own house, sometimes helping out and just as often getting a little of Mittie's home cooking. I was one of those kids who could always eat more than my share. Unfortunately, some things learned in childhood never leave you. She always had a big bottle of soda pop for me and a peppermint stick or a pork chop. I don't have a clear memory of watching Mittie cook lunch or dinner, but I could never forget Granny's breakfasts. I still remember the sound of the ham steaks or pork chops sizzling in the old iron frying pan with a couple of eggs bubbling on the side. The memory of her hot biscuits dripping with butter and homemade grape or plum jelly still makes my mouth water.

Granny's breakfasts were the best part of getting to stay the night, which I sometimes did. She lived in a small white frame house that looked out on the lake, and the two of us would sit on the front porch in a pair of rockers and watch the world go by or at least what there was of it in Eufaula. Or we'd stroll down to the lake and put a couple of poles in the water with a cork and a hook and stretch out on the green banks waiting for "dinner" to arrive. Sometimes she'd talk my head off, and other times we'd sit quietly for hours. She taught me a lot about life, and when the fish weren't biting, I learned at least as much about patience. That's a talent I find comes in handy on a regular basis in Washington.

When I got older, 9 or 10, I'd go down to Mittie's and do chores for her—mow the lawn, dig up worms for fishing. At our house or Mittie's, when you got old enough to put your fingers around the handle of a mower or a shovel, you were put to work.

Mittie's life, like my mother's, wasn't easy, but it embodied the values of sacrifice and hard work. Mittie and some of the other older women in the neighborhood were still out in the fields picking cotton well into their sixties. These unbelievable women would throw a 14-foot sack across their backs, fill it full of new cotton, and drag their load to the scales for weighing. That was backbreaking work for any youngster, let alone a group of grandmothers. My brother Lawrence got it right when he said, "They sure was makin' ladies back during that time!"

Both my mother and my grandmother taught me that there is an innate virtue and characteristic that women possess when it comes to their children—a special bond that defines the relationship between mother and child. I see it in my own family today with my wife, Frankie, and our children. Don't get me wrong. I think fathers are extremely important to every child and play a crucial role in a child's moral, mental, and physical development. I've got a great relationship with my kids, but when it comes to the nurturing authority and love a mother brings, dads aren't in the same league.

Mittie died on a hot summer morning in June 1983. Just a few days before, I had been near Eufaula giving a speech, and I passed through my hometown on the way back to Norman, where I live today. As I breezed through town, I thought to myself, "I don't have the time to go see Granny, but I'll catch her next trip."

Three days later, I was home mowing the lawn and feeling a little guilty that I hadn't stopped to see her when my sister Gwen called to tell me that Mittie had died that morning. I think we all get our priorities confused from time to time. I know Mittie would have understood that I wanted to get home to my own brood that day I didn't stop, that I was trying to make my mark on the world, to establish myself and provide for my family. But that didn't make me feel any less guilty.

Mittie was as independent as they come and didn't expect or want us to feel obligated to her, but I was beholden to Granny for more than I could ever repay.

There are moments in life that you want to take back, to do over again the right way. Mittie was gone, and I had passed up the opportunity for one last afternoon with Granny.

When my mother died in 1993 after a long struggle with diabetes, I felt much the same way. She had been my port in the storm for 36 years. When I was a kid, she would sometimes threaten me with "I'm going to tell Buddy," and other times she was my life-saving defender when I found myself in Buddy's crosshairs. Later, she was a wonderful grandmother to my children, giving them the same tough love she gave me.

Before both Mama and Mittie died, I think I somehow knew deep down that God would soon call them home, but that didn't make their passing any easier. There's an old saying: "You can be prepared, but you're never ready." I don't think anyone is ready when a loved one is taken, but the hole left in your heart forces you to take an inventory of your life and reflect on what's really important. Their deaths reminded me that no matter how far you go in life, someone else helped you get there.

Mama and Mittie were the stabilizing forces that guided me on the right path—not necessarily the same path my brothers and sisters—or most of my friends or fellow African Americans—would take. How I feel about the importance of strengthening our families, of giving children both boundaries and stability, of teaching that real wealth and true dignity cannot be measured by the size of our bank accounts but only by the depth of our hearts, I owe to these two strong women. Together, they left me a rich legacy—their love and values, a safe place from which to fly as far as my dreams and determination would take me, and a gentle push to send me on my way. Not a day goes by that I don't think of Mama and Mittie and thank God for every day I spent with them.

4

God Gives You What You Need; You Have to Work for What You Want

Personal responsibility is the brick and mortar
of power.

> Shelby Steele
> *The Content of Our Character*

If I had a nickel for every person I've met since elected to Congress in 1994, I'd be a rich man today. One thing I've learned is that the people who come in and out of your life give it texture and meaning. Republicans, Democrats, supporters, opponents, reporters, and people you just meet along the way all keep you humble, help you do the right thing, and once in a while even give you a pat on the back. For me, the opportunity to meet people from all walks of life and from every part of the country is one of the things I've liked best about being a member of Congress. But anyone who has spent time in Washington knows that a lot of the folks who populate this town have about as much in common with the rest of the country as Jesse Helms and Jesse Jackson on a good day.

It's a different world "beyond the Beltway," as they say in the nation's capital. Outside Washington, people are a whole lot more concerned about a late Social Security check than the latest political scandal. They want to know what you're going to do to help their kids learn to read and write, not whether you *lean* to the right or the left. Leave the D.C. city limits, and most people take you at face value. They listen to what you have to say, they don't question your motives, and they give you the benefit of the doubt until you prove them wrong. I've tried not to do that very often.

The value system in Eufaula was much the same. Your word mattered a lot more than how much money you had in your pocket. Your honesty was more important than the house you lived in or the car you drove. Unlike in Washington, it wasn't who you knew that determined how far you could go, but how hard you were willing to work. In my family, we learned early on that personal responsibility was the only direct route to the "promised land," and you didn't get respect unless you earned it.

Helen and Buddy Watts and Grandma Mittie gave me a lot of things, but they never gave me an excuse to fail. They gave me a strong work ethic. They taught me a lot about integrity. I understood at a very young age that each of us is responsible for our actions and ourselves. Nothing in the last 44 years has led me to doubt the wisdom of what I learned from a family that often faced and overcame hardship.

I believe in these values—not just for any one ethnic group, but for our nation as a whole. They guide us to a better place for ourselves and our families. Of course poverty puts people at a disadvantage—black or white. Add racism to the mix and the odds get longer, but they're never insurmountable. My father is a perfect example. By all accounts, Buddy Watts lived a successful life. He was never rich, not even close to it, but at his death, he was financially okay, respected in the community, and loved by his family. Most of us will thank the good Lord if we end up half as contented as my daddy when Heaven calls. But I've always believed that if Buddy had been born a couple of generations later, when the promise of equal opportunity that he fought for had become a reality, his talent for business would have taken him to the top.

Back in the mid-1970s when I was sporting an Afro and about to go off to college, Jesse Jackson was inspiring young African Americans like me and my brothers and sisters by telling us, "You *are* somebody." He was right, and so was that message, then and now. "I am somebody" is as truthful and potent a statement for my kids today as it was for me and my buddies in 1975 growing up in Eufaula. Ten years earlier, the Civil Rights Act of 1964 had been a cultural earthquake in our country, but while Jim Crow may have been officially stricken from the law books, its legacy remained in too many hearts. A decade after Lyndon Johnson signed that historic act, we were still trying to build a society of equal opportunity from the rubble of Jim Crow racism. Young black Americans desperately needed to believe in themselves and their ability to achieve their dreams, and not everyone was lucky enough to have the kind of family I had to instill in me the values of self-determination, hard work, and personal responsibility. Reverend Jackson challenged young people to reject the second-class status of their parents' and grandparents' generations, and thousands of black children answered affirmatively, "I *am* somebody."

But over the past 25 years, this proud statement has been traded for a hollow rhetoric that is nothing more than a sad and shallow excuse for failure and misbehavior. Despite the enormous progress of the African American community by almost every measure, "I am somebody" today has become "I am somebody's *victim.*" I reject this fashionable "cult of victimology," as Professor John McWhorter calls it in his book *Losing the Race.* My upbringing simply does not allow me to take the easy but inevitably defeatist road to victimhood so well worn today from over-use. I can remember hearing my parents and grandmother talk about race and discrimination. They weren't blind to the world around them, but never once did any one of them say, "I can't because I'm black."

Sure, our lives were made more difficult in those days because we lived on what some people thought was the "wrong" side of the tracks in Eufaula. Money was in short supply. There were no extra luxuries in our lives because Daddy was always servicing the debts on his properties. And because he'd grown up during the Depression, paying off his notes ahead of schedule was a point of honor for him.

I didn't have the clothes or cars that kids have today, so by those standards, I suppose my life would probably be judged as lacking. But nothing would be further from the truth.

Marian Anderson, a woman of enormous talent and courage, faced many of the same kinds of difficulties I did growing up in a poor black family. Reflecting on her childhood, the great diva said, "It is easy to look back self-indulgently, feeling pleasantly sorry for oneself and saying I didn't have this and I didn't have that," she once wrote. "But it is only the grown woman regretting the hardships of a little girl who never thought they were hardships at all. . . . She had the things that mattered."

So did I. I was raised in a strong and loving American family with a mother and father of solid moral values and an extended family of terrific role models. My roots, which I'll always treasure, have sustained me every step of my journey, and the success I have achieved had its genesis in Eufaula, Oklahoma. Growing up there gave me concrete ideas about work that drove me to get my first "non–Watts family" job at 12, play amateur and professional football, go to college, enter the ministry, and hold public office.

My family and my community gave me a good understanding, some good bearings to get out in the world and try to navigate my way. It wasn't a part of my training to be a victim. That wasn't in the manual. My parents didn't raise victims.

I was somebody. I was a Watts.

And at the Watts house, you didn't have time to lie around and feel sorry for yourself, much less freeload off Daddy and Mama. If you lived under Buddy Watts's roof, chores were a fact of life. Nobody in my family ever spent Saturday morning in bed. The boys would be out working with Daddy by daybreak, and the girls spent the weekends helping Mama clean house. And if you were working outside the family and not in school, you helped pay your own way by chipping in a little bit for groceries and the mortgage.

Those rules went for every one of Buddy and Helen Watts's kids. Life in the Watts family was no Sunday afternoon picnic, but hard work never hurt any child. And all of us managed to live through our childhoods none the worse for wear.

"Upscale" isn't a word used to describe the housing on either side of the tracks in Eufaula very often. For the most part, it's a town of simple homes and plain streets that slope gently down toward the lake. Once you cross the tracks, the houses get smaller and more of them need a coat of paint or Daddy's kind of magic, which could turn a broken-down shack into a home.

We grew up in a series of my father's rental houses that were acquired under the Buddy Watts "four steps to prosperity" plan.

- Scrape together a down payment on a run-down house.
- Use a lot of elbow grease to fix it up—paint it, put in a new floor, or even add on another room.
- Move the Watts family in until the next house comes along.
- Rent the house and pack up for the next move.

Daddy's "gentrification" business was a family affair. All of us kids helped with the renovations while Mama settled us in and put her touch to the place. Then a few years later, the whole process would start all over again.

For a while we lived right next door to Grandma Mittie. This house was a single-story, 700- to 800-square-foot house. Then we moved to another, this one two stories high, about 50 yards from the church. All of us kids slept together in one big dormer room—boys on one side, girls on the other. That house holds a special place in my heart, not only because it's where I had my first birthday party, when I turned 7, but it was at that party in that house where my future wife, Frankie, caught my eye.

When I was in second grade, Daddy had a house built, maybe 1,200 square feet, again a single story on a corner lot. We thought we'd died and gone to heaven. Imagine—1,200 square feet for six people—this was spacious living for us. The mortgage was $88 a month—a lot of money for Daddy to come up with in those days. When times were hard, Daddy used to threaten to "let the bank take it!" But we'd hold on, tighten our belts, and times would always get better.

When I was in seventh grade Daddy bought the "big house." It was about 2,000 square feet, two stories high, a corner lot. This really was a mansion by our standards, especially since there were only five of us living under the Watts roof by then.

After I went off to college, Buddy tore that "mansion" down and built the brick home he died in. He was so proud of that house. Where we come from, a brick home kind of says that you've arrived. There weren't many in our neighborhood, and there still aren't.

With six kids, space was always at a premium, but we learned, like any big family does, to adjust to the room we had and be glad to have a roof over our heads. All in all, we didn't mind moving as long as it was in Eufaula. When we were young, the family had spent too much time on the road as Buddy looked for work, so we were happy to have some permanent roots at last, even if we transplanted ourselves down the block or across the street from time to time.

We moved so often over the years, it's a wonder we got any of our mail. But that's another advantage of a small town: the postmaster always knows where to find you. Of course, so do the sheriff and the bill collector.

But the law didn't look for the Watts kids very often. We weren't angels—far from it—but when we were up to no good, at least we were pretty good at not getting caught. My brothers Lawrence and Melvin weren't serious troublemakers, but they were an independent pair who could get themselves in dutch with Daddy every now and again.

They still love to tell the story of the Christmas they outsmarted "Santa Claus."

That year, Buddy bought the two of them new bikes and hid them behind the smokehouse awaiting Santa's "arrival." Daddy should have known that it wouldn't take Lawrence and Melvin long to sniff out those bikes. They spent the week before Christmas riding them all over Eufaula in the afternoons and putting them back before Buddy got home from work. It wasn't exactly the stuff of Pretty Boy Floyd, but they got a big charge out of pulling the wool over Daddy's eyes. That didn't happen very often.

Today, I'm much closer to my brothers than I was as a young boy. When we were kids, too many years separated us to be real buddies. My older brothers were close in age, and they sort of formed their own club. The only thing missing was a sign that said, "No Baby Brothers Allowed!" That meant me.

Melvin was the typical oldest child—the weight of the world fell on his shoulders. When Daddy was out working, Melvin was expected to fill his big shoes, and Mama would put Melvin in charge of the rest of us if she was gone from the house. Much as we sometimes chafed under his command, he did a pretty good job of keeping us out of trouble.

Melvin was a good football player. My high school coach, Paul Bell, would often talk about what a good running back he was. But like so many in Eufaula back then, he quit school, moved out of the house and in with Mittie when he was only 15. By the time he was 17, Melvin was married and on his own. In that way, he was a lot like Daddy.

Both my parents were fairly strong-willed people, and I think all of us inherited at least some of their stubbornness, but Melvin, Mildred, and Lawrence may have gotten a slightly bigger share. Melvin had a hard time with the restrictions of the Watts household.

Daddy used to say, "If you're going to live under my roof, you're going to live by my rules." Melvin took him at his word and moved out.

Lawrence was just as willful, but much more quiet about it. Of all the Watts kids, Lawrence reminds me most of my mother. He doesn't say a lot, but when he does, you know he means it. Mama was a woman of strong opinions, too, who watched over her flock like a mother hen. In some ways, Lawrence is the same way. He's always there when you need a friend, someone who will listen to your problems. It was Lawrence and my sister Mildred who spent many a night sleeping in a chair next to Daddy's bed when he was sick in the hospital.

He's also one of those guys who pops up when you least expect it. When I was in college, I wouldn't hear from him for weeks at a time. Then, he would show up at two o'clock in the morning, knocking on my door, wanting a place to hang his hat for the night.

Daddy used to complain about what handfuls my older brothers were, but Lawrence didn't push the limits as much as Melvin. He liked his warm bed and Mama's home cooking too much to risk Melvin's fate. Lawrence is one of the hardest working people I know, and always has been. He finished high school and was offered several football scholarships, but he turned them down. School just wasn't a big deal to him. By the time he reached his senior year, he wanted his independence, and maybe he thought college would delay that.

Instead, he headed off to Oklahoma City and Wichita after graduation to find work. Before long, he'd saved enough to open a little nightclub in Eufaula. It didn't take Lawrence long to discover that the profits weren't worth the problems that came with heavy-drinking customers. He traded his barkeep's apron for a bricklayer's trowel. He and Melvin both became jacks-of-all-trades, much as Daddy was. Whether it was farming, ranching, carpentry, plumbing, or hauling, the Watts brothers could do it all and did.

As much trouble as they gave him when they were growing up, my brothers both followed a course much like Daddy's.

Next in line came my sisters Mildred and Gwen. I spent a lot of time with them when I was a small boy. Just like her older brothers, Mildred had her strong opinions. She always stood her ground pretty effectively, too. Gwen wasn't shy either, but like Lawrence, she exercised a little more patience and diplomacy. Today, Mildred is married and the mother of three children. Not long ago, she retired from a long career with Southwestern Bell after working her way up from operator to management.

Gwen, my middle sister, ended up being the first Watts to graduate from college. Like all the Watts kids, she worked very hard. She was always a good student, and now she's a special education teacher. I always thought Gwen was special, and now she's got the title to back it up.

My sister Darlene is as much younger than me as I am to Gwen. Mama and Daddy relaxed the rules for her, being the baby, and she grew up after most of the older kids had left home. I think my parents looked the other way, maybe more than they should have, when Darlene strayed from the right path. She was always a free spirit. Nevertheless, she is cut from the same cloth as the rest of us and today works in the health care field.

As we grow older, the difference in years between siblings matters less and less. That has certainly been true for me. My relationships with my brothers and sisters are quite different today than they were when we were children. I love them all and treasure the closeness I think we all feel for each other. Part of my feelings toward them comes from knowing how much each of them did for me as a child, but as a young boy, I felt a special bond with Gwen that I can trace back to a single moment on a playground nearly 37 years ago.

It is one of the experiences I remember well from the years when we uprooted ourselves every winter. I was attending Dunbar Elementary, a segregated school in Wichita, Kansas. We were living in a housing project several blocks from the school, and I was miserable. I missed my buddies back home in Eufaula, and the winter was brutally cold that year, colder than I had ever known winter could be. There was no bus service for us, so Gwen, Lawrence, Mildred, and I would bundle up and head off to school each day wishing God would shine a little of his warm light our way.

The cold and snow came in icy blasts, and I was too small to walk easily in that terrible winter wind. Soon I'd be crying, and it was Gwen who would get me home to Mama, where they would both rub my hands for warmth.

But it was on the playground that Gwen was my real protector. She was in the sixth grade when a boy in her class decided I would make a great punching bag. During recess one day, he singled me out and pushed me down to the ground. I don't think I was really hurt, probably just making a little noise for sympathy's sake, but Gwen lit into that boy like a fury. It was a real throw-down fight, and she cleaned the playground with him. I loved it! That man must be middle-aged by now, but I bet he can still feel the whipping he took that afternoon from Gwen Watts.

When we got home, Gwen had some explaining to do to Mama about her torn clothes.

"I had to take care of my little brother," she told the folks that night. Now, as a teacher, she's always on the peacekeeping side of playground disputes. But she still gets a big smile out of remembering the day she came to my rescue. All I knew back then was that I had a big sister who could fight!

That was the beginning of a great friendship that continues today. Gwen and I also teamed up musically during my grade-school years. We weren't exactly the Jackson Two and certainly not Ike and Tina, but we had a lot of fun entertaining our elders. Gwen and Mildred are both pianists, mostly self-taught, and by the time Gwen was in high school, she had become song leader in our church choir. That was the Sulphur Springs Baptist Church, a low-roofed building where the air-conditioning was provided by the congregation waving paper fans printed with inspirational messages and a local business ad.

Gwen would play, and I would sing. If I'd been center stage at Carnegie Hall, I couldn't have been any happier than I was singing to the folks of the Sulphur Springs Baptist Church on Sunday mornings. I suppose the passage of time puts a convenient golden glow on our recollections. I know we *thought* we made the congregation rock! But I did hear people say, "Look at that little boy. He's going to be a preacher." Shyness wasn't in my nature, and belting out hymns like "Another Day's Journey" and "That's All Right" was more fun for me than putting balloons on my bike to make noise and then terrorizing the neighborhood—well, almost.

I've always been grateful to Gwen and Mildred for introducing me to the joy of music that became even more important to me as I grew up. In fact, at one point, I dreamed of becoming the next Al Green or Marvin Gaye. My life took a different turn, but like Walter Mitty, there was always a little voice inside me wondering, "What might have been?" At the 2000 Republican Convention in Philadelphia, I finally got a chance to find out. Some good folks held a reception in my honor on the first night of the convention with some very special guests, the Temptations, one of the truly great Motown groups. It was a fun evening for everybody, especially yours truly. With hundreds of friends and supporters looking on, I finally got the chance to sing lead for a rousing rendition of "My Girl" with a group that has long been one of my favorites. I didn't get a record deal out of the appearance or an invitation to tour, but I don't think I embarrassed myself either. In fact, E. J. Dionne, a columnist for the *Washington Post,* said I "merrily belted a rather good version." Not exactly a rave review, but I was grateful. It was the most positive comment about a conservative I'd seen E. J. write in recent memory. That evening was one of the high points of my life. I'll never forget it and neither will my kids.

When we were children, Gwen was always on the go. She was a popular cheerleader, like Mildred, who led a fairly typical teenage life with studies and boyfriends to keep her busy. And like my other brothers and sisters, she held down a part-time job. The last thing she needed was a small boy tagging along, but she and Mildred often kept track of me for Mama.

In *Peanuts* Linus says older sisters are the "crabgrass in the lawn of life." In my case, I was the pesky weed they couldn't get rid of. Gwen's boyfriend in high school was Lucious Selmon, who later became famous as a star defensive lineman for the Sooners, but in those days, he was just the guy I saw trying to hold hands with Gwen on the old green couch in our living room without getting caught by Mama or Daddy—no small feat. I would always ask him if I could feel his muscles. He'd grin and curl a forearm back. When his bicep stood up, my two hands weren't big enough to wrap all the way around it.

On Saturday nights, Gwen and Lucious would go out after she got off work. When Gwen or Mildred wanted to go out on a date, whether with Lucious or anyone else, I was always sent along as "chaperone." This was Mama's way of making sure her daughters wouldn't be tempted, like most teenagers are, to kiss in the movies or anywhere else. Date or no date, Helen Watts would not allow her girls to act like "fast heifers," as she called it. What Mama never knew was that I had my own little protection racket going. Gwen and Mildred paid me to get lost for a couple of hours and then hook up with them just in time to arrive home together. The girls got some quality time with their boyfriends, I got "rich," and Mama got snookered—one of the few times we got the better of Helen Watts.

With the cost of raising children going up and up, the trend today is toward smaller families. I took another path here, too. Frankie and I have five children. We know a big family is both a serious responsibilty and a wonderful gift. There may be less money to go around; but there are more people to love and take care of, more competition to give them gumption, and more role models to help them find their way. That's how I saw my brothers and sisters.

Each of the Watts kids was a very different individual, but we all shared a love for each other and a deep-held belief in the Watts family creed: God gives you what you need; you have to work for what you want.

Digging in and doing my part as a kid made me feel that I was contributing to something bigger and better than self-interest, and I found out I liked the feeling. It was a powerful lesson that I believe led me later in life to the ministry and then to politics. Working at a young age also helped my confidence grow in my own ability to succeed.

I never had an allowance. Daddy didn't believe in it, although my parents would occasionally give me some spending money. During my junior high years, I decided this arrangement needed revisiting. I started thinking about the $35 that rolled in every month from one of the rental houses Daddy had bought and refurbished near us. I figured a $35-a-month allowance would translate into a neat $1.15 a day, just floating in the door with no work involved! If you knew what to do with it, that kind of money could go a long way. Mr. Sidney Shavers, who owned one of the corner stores nearby, would cut off about half an inch of bologna from his deli case for just a nickel. A dime's worth was almost more than I could eat—although I usually managed. A little cardboard container of chocolate milk was just 5 cents in the school cafeteria. I figured I could feed half my friends on this allowance and still have money left over.

Now I knew Buddy wasn't going to fork over the whole $35 to me and with the benefit of a little experience, I also knew there was safety in numbers when it came to dealing with Daddy. I thought that if there were two of us, it would be a lot harder for him to turn me down. Mildred had graduated by then, but by proposing a 50-50 split, I lured Gwen into this high-stakes capital venture to convince Daddy of the merits of my plan. We knew it was going to be a tough sell. We decided to tell him that giving the $35 to us directly would be kind of like an allowance. Until then, the only kid we'd ever seen get an allowance was Beaver Cleaver on television.

But we were undaunted by the difficulty of the task. We cornered Daddy in the living room one afternoon and spelled out the proposal as convincingly as possible. We were mature and logical. We were enthusiastic and convincing. We were doomed.

"If I gave you our rent money," he explained in a steady, quiet tone that let me know the discussion was over, "you wouldn't hustle."

And he was right: Free money would dampen our interest in watching out for side-of-the-road pop bottles or canvassing the neighborhoods for lawn-mowing, pecan-picking, and hay-hauling opportunities. We were pretty disappointed for a couple of days, but then it was back to reality, which wasn't so bad. Every time I could round up a little work and end up with a little spending money, I always felt good about myself and what tomorrow would bring.

My hustling days began pretty young, just as it had for my older brothers and sisters. Lawrence can still remember going along to pick cotton as a boy on the first day of picking season. Picking cotton was hard work that usually left you with torn hands and an aching back.

Most of the kids would be off playing long before they'd earned any real money, and by the next day, their numbers would always dwindle. But Mittie made Lawrence keep up with her pace. It wasn't long before he was earning $6 a day or even more. With a new pair of jeans costing less than $2, Lawrence quickly learned that when the cotton was high, so were the earnings if you didn't mind the work.

My own introduction to the joy of manual labor began at the ripe old age of 7 working alongside my father and my brothers in the hay-fields. Daddy would contract out with farmers to haul their hay. It was hot, scratchy work, but from the first day, I was treated like part of the crew. I may have been the smallest, but I felt more than halfway on the road to becoming a man, especially sitting behind the wheel of a 2-ton GMC truck.

My feet couldn't reach the gas pedal, but that didn't matter. Edison would have been proud of my daddy's ingenuity in rigging the truck for a 7-year-old. He'd put it in what's called compound gear—the gear you use to uproot an old tree stump. I'd climb up into the driver's seat, a definite reach for my short legs, but I usually made it in one try. Daddy would get the wheels headed straight down the row, leaving the engine at idle. The truck would roll just slowly enough for my big brothers to toss the hay bales up on back while Daddy stacked them. My job was to steer the truck straight down the hay rows, which sometimes seemed 10 miles long.

When we got to the end, Lawrence or Melvin would hop aboard, turn the truck around, aim it down the next row, and then turn the wheel back over to me. Year after year, I kept trying to reach that gas pedal, though, and finally one day, I discovered I was long-legged enough to give it a solid nudge with my toe. The big truck lurched forward, and hay bales flew every which way.

"Boy!" Daddy yelled while Melvin and Lawrence cackled, "What're you doing? Don't you ever do that again." He didn't have to worry. I think I scared myself more than the rest of the crew. Once I'd hit the gas and felt that 2 tons of rubber and steel bucking almost out of control, I truly didn't want to do it again. I just wanted to know what it felt like, and now I knew.

Once my need for speed had been satisfied, being big enough to lift a bale of hay was my next ambition in life. Looking back, it's obvious that I hadn't reached the age of reason, but getting that first bale onto the back of a truck was a rite of passage to manhood in the Watts family. A bale of hay weighed about 75 pounds, though many were bigger. I kept working at it and working at it and finally got to the point I could lift and heave one of those bales as high as the truck bed by myself. I was practically a man! And then I realized—too late—that I liked driving the truck a whole lot better.

By the time I was 8, I'd begun helping Daddy fix up the beaten-down houses he usually picked up for a song and converted through sweat equity into decent rental homes. My job was to pull bent, rusty nails out of reusable old boards in those houses and stack the boards by size. Daddy paid me a penny apiece for every nail pulled. Whenever I got the urge to quit, I would mentally picture a trickle of pennies, one after another, dropping into a mason jar, and I went back to work again. The way I'd pile those brown nails up so carefully, you'd have thought I was stacking gold bars at Fort Knox. But, I learned that every penny counts, and to this day, it's hard for me to pass a penny anywhere and not pick it up. I learned the value of a penny early in life.

When we were kids, Gwen, like our brothers and Mildred, always held down a job, too. Because we were so close in age, she was probably my most important influence when it came to understanding the value of personal responsibility and hard work. She helped me get my first job away from family-based enterprises, and she was so good at saving money that she was able to buy her own bedroom set and stereo. Her financial independence impressed me in a way that all Daddy's money lectures in the world couldn't do. She was the example I wanted to follow. The importance of putting some money away for a larger purpose was yet another lesson learned from my family that I never forgot.

I got my first "real" job at age 12—clearing tables and washing dishes at JM's Café for 65 cents an hour. Gwen was already working there, and she was such a dynamo that Betty and J. M. Bailey, the owners, were willing to take a chance on me.

Their café is still about the liveliest one in town, almost always busy. It sits on Selmon Road, eastbound into town, a half-block from Main Street. The Baileys started their business in 1955, two years before I was born. It hasn't changed much since the days I was clearing tables. Wander through the front door, and you'll run right into a glass display case full of big, fresh steaks to choose from. My mouth used to water just staring at that prime beef—still my favorite meal. And if a steak dinner doesn't fill you up, there's another case crowded with the most luscious cream and fruit pies you'll ever eat—one of my biggest weaknesses, as anybody in Eufaula who knows me will tell you.

Even now, a breakfast at JM's with two eggs, thick ham steak, and all the trimmings costs less than five bucks. Going to IHOP just isn't the same once you've been to JM's. I suppose Weight Watchers might frown on a JM breakfast, but I guarantee it'll keep you going till the last bale of hay is on the truck.

J. M. and Betty Bailey have probably employed more black folks than all the other businesses in Eufaula combined. In fact, Mittie even worked for them at one time. Occasionally, they'd stick their necks out and nurture some rougher character, the kind who might be prone to getting drunk and missing work every so often. But most of all they offered hardworking black people a much-needed chance: an honest day's pay for an honest day's work.

Sometimes, when the action got heavy at JM's I even helped cook hamburgers and french fries. I was only 13 and pretty excited about that level of responsibility. Weekdays, I had about half an hour of free time after my classes got out. Then, from 4 until 10, I earned $3.90, went home, and dropped into bed. My Saturday duties were from 2 to 10, which got me more than $5, but sometimes it was tough to see my friends running off to swim or play football or baseball on the church grounds, especially in the summer. Still, I loved my good supply of spending money, and if Daddy wasn't going to give me his rental money, and he wasn't, then I decided to get an allowance the old-fashioned way. I'd earn it myself.

The two and a half years I worked for the Baileys also gave me great respect for the employer-employee relationship, and my strong belief in the importance of small business to the U.S. economy and its families. The Baileys are typical small-business owners, willing to go that extra mile for you. They recognized the importance of investing in people, and you could count on J. M. to give you a fair deal. At JM's, you didn't need a labor contract, just a handshake. If you were a good worker, they'd always be good to work for. That mutual trust and respect, learned at a young age, reinforced what my own parents taught me about personal responsibility and earning the respect of others.

But if my parents gave us room to grow, they also helped us learn how to fail, which is an important part of growing up, too. The money we earned outside the family was ours to spend, wisely or foolishly. That was great psychology. Few things can give a youngster such a sense of accomplishment and empowerment as buying something with his or her own money.

But as much as I enjoyed spending my paycheck, I was to learn quickly that overspending had its definite drawbacks. Chenault's, the top clothing store in Eufaula, was on Main Street, not more than a hundred strides around the corner from JM's. Robert Rinehart was the manager, and he came in every day for a late-afternoon supper. One day, he called me over to his spot at the counter. "J. C.," he said, "I've had an eye on you for some time." That was true; I'd been there a whole year. "You're such a hard worker," he went on, "that I'm going to let you buy things on credit at my store any time you like!"

An alarm bell should have gone off in my head, but ego triumphed over common sense. I forgot everything my daddy and mama ever taught me about money in that moment as a vision of ol' J. C. in a cool new outfit flashed before my eyes. The very next afternoon, I strutted home showing off a brand-new sports jacket and beautifully coordinated slacks. I thought I looked fantastic! Mama agreed, but her eyes said something different, and it wasn't good.

"So what," I thought. "It's my money anyway." Then I got an economic reality check. At 65 cents an hour, $90 was a mighty, looming debt to service. By my calculations, I was going to be sweeping, washing, and frying for almost 140 hours to pay for that outfit. Suddenly, looking good didn't seem nearly as important as getting out of debt.

I learned you can't escape mathematics, a fact that *has* escaped more than a few people in Washington. That outfit ate up almost every cent I could earn at JM's in the span of a month! It was my first and last splurge at Chenault's. When I told Daddy about my experience in deficit spending, he smiled at me and said, "Junior, whenever your outgo exceeds your income, your uplift will come to a downfall."

After my stint at JM's, I became a jack-of-all-trades when it came to making money. As a teenager, I worked at the state lodge nearby, hauled hay, mowed lawns, delivered newspapers, and did almost any odd job that came along—always hustling for a little cash. Back then, a quarter went a long way.

On hot afternoons, my buddies and I would walk to the Tastee Freeze out on the north end of Main Street. The townspeople would park under the long shade roof of the carhop area while the rest of us swarmed up to the order window. With money in my pocket, I was proud to be able to buy ice-cream cones for everyone in my pack.

They say kids often take on the traits of their parents. I believe you live what you learn. When it comes to pulling yourself out of poverty through hard work, I've lived it through my family, and that experience formed the value system that has guided me in Washington. Seeing my daddy hold down three and four jobs at a time, while all my brothers and sisters worked and went to school, taught me the value of personal responsibility and influenced my decision in 1995 to go against the members of the Congressional Black Caucus and support welfare reform. I saw firsthand that with the right attitude and effort, hurdles— even high hurdles—can't keep you from success if you want it badly enough. That doesn't mean people don't sometimes need a helping hand getting out of the starting blocks.

The Community Renewal Act, legislation I was proud to co-author and pass into law, came from the roots of my raising in Eufaula when I saw my hometown, a poor rural community like so many, struggle to provide economic opportunities for its people. But this is no government giveaway. Instead, through innovative incentives, it empowers low-income Americans to lift the weight of hopelessness, rise up, and achieve their dreams on their own.

It wasn't a political party's platform that convinced me to raise questions about the notion of preferences and set-asides, which riles a lot of national black leaders. Ironically, that attitude came straight out of everyday life in a cohesive black American family. And, as you will read later, that same upbringing was behind my decision to caution my party against ending affirmative action prematurely. My roots also taught me that working people deserve lower taxes, and that's why I opposed President Clinton's tax increases and enthusiastically supported President George W. Bush's tax cuts. When it comes to tax relief, the only color that matters is green.

For those who can't understand how I came by my political views, they need look no further than a little town called Eufaula. I think it's fair to say rural America produces more than the food we put on the table. It raises children with dreams and the values to reach them, and thankfully, there are Eufaulas sprinkled across every state in the Union.

One of those lies 270 miles north of my hometown as the crow flies. A century ago, Abilene, Kansas, was known as the last stop on the Chisholm Trail, much as the Eufaula area was the final destination of many who walked the Trail of Tears. It was a cow town with a rowdy beginning, like ours, that produced one of this country's greatest heroes: Dwight David Eisenhower.

Head west across the Kansas flatland about a hundred miles, and you'll run right into another small town that grew up next to the Union Pacific tracks. Russell, Kansas, too, can claim a homegrown hero, and one of the most respected men in Washington today: Robert Dole.

Most people might wonder what the thirty-fourth president of the United States and the former majority leader of the U.S. Senate all have in common with the grandson of a poor black sharecropper.

I certainly don't put myself in the same category as Dwight Eisenhower or Bob Dole. Men like these don't come along very often, and we can only thank the Lord for putting them on this earth and giving us heroes that make us strive to be better. But I am proud to share something very important with these two extraordinary men—the values of faith, family, personal responsibility, and hard work that came from growing up in a large family and a small town. Dwight Eisenhower came from a family of six brothers and not a lot more. He once wrote of his childhood, "We had been poor, but one of the glories of America, at the time, was that we didn't know it. It was a good, secure, small-town life, and that we wanted for luxuries didn't occur to any of us."

When I first read those words, actually in Bob and Elizabeth Dole's wonderful book, *Unlimited Partners,* my first thought was how much I would have liked to meet and talk with President Eisenhower. I'm sure he would have understood why I feel so strongly about the importance of personal responsibility. He lived it, too. He knew what hard work was. Ike stacked hay with his brothers just as I did with mine. They plowed the fields and pumped water by hand. He once described his boyhood home in this way: "It provided both a healthy outdoor existence and a need to work," he wrote, "and any youngster who has the opportunity to spend his early youth in an enlightened rural area has been favored by fortune." Amen!

President Eisenhower died when I was still a very young child, too young to understand his contributions to war and peace and the cause of equal opportunity. Luckily, I have had the chance to get to know Bob Dole, and each time we've talked, I've been reminded of his deep commitment to family values—a commitment that he and my daddy shared.

Like Buddy Watts, Bob Dole came of age during the Depression and faced tough times as a child when money was short and his parents struggled to feed their children. He's never forgotten his roots any more than my father could have forgotten his, and it has earned him the reputation as one of the finest men ever to grace the U.S. Congress.

The lives of Dwight Eisenhower and Bob Dole and Buddy Watts all teach us that nothing—not poverty, not racism, nor physical disability—need keep us from our dreams. In the United States, we *can* grow up to be president, or majority leader of the Senate, or representative of the Fourth Congressional District of Oklahoma. God *does* give us what we need, but each of us must work for what we want.

I don't get home to Eufaula as often as I'd like anymore. While I've been back in the district almost every weekend, Congressional duties and three growing children still at home have kept me pretty busy. But Melvin, Lawrence, Mildred, Gwen, and Darlene are never far from my mind and heart. On Monday or Tuesday morning, when I've had to catch the early flight to Washington, I sometimes struggle to get out of bed, and I drift off for just a couple more minutes. Suddenly, I find myself back in Eufaula. It's early on a Saturday morning, and I can hear Daddy hollering at us, "You boys gonna sleep all day?" And I smile.

Then, I remember Mittie in the cotton fields or I think of Mama's aching hands from wringing clothes all day. I see my brothers hauling hay in the hot sun or my sisters washing dishes at JM's or cleaning house under Mama's watchful eye, and I get myself out of bed, glad to be alive and grateful to every one of them for what they've given to me.

5

If You Wait for the World to Change, You'll Never Change the World

We live under the same sky, but we don't all have the same horizon.

Konrad Adenauer

I've never known anybody to achieve anything without overcoming adversity.

Lou Holtz

M ine wasn't a Norman Rockwell childhood. There were too many black faces, too much poverty, and too little opportunity for it to look like a Rockwell portrait of America. Back in the 1950s and 1960s, the depiction of everyday life in a black community wasn't the "right stuff" for the cover of the *Saturday Evening Post* or for most magazines of the time. In those days, the media rarely portrayed African Americans in anything other than stereotypical subservient positions—as maids or porters, stable hands and janitors.

After nearly a half century with the *Post,* Rockwell did his last illustration in 1963. But a few months later, on January 14, 1964, the readers of the *Post*'s competitor, *Look* magazine, got a surprise—a full-page Rockwell illustration. However, instead of his usual celebration of small-town life that people had come to expect, they saw something quite different—a small black girl in a starched white dress, her arms filled with books, marching bravely off to school escorted by four federal marshals. Norman Rockwell's "The Problem We All Live With" struck an emotional chord in people of every color as it told another kind of American story—the defeat of shameful segregation through the courage and determination of a black minority.

In 1960, four years earlier, little Ruby Bridges walked into William Frantz Elementary School in New Orleans and made history by becoming the first African American child to attend an all-white public elementary school. For an entire year, she was the only child in the school, as white parents kept their children home in protest, but each day little Ruby made that trip to school the country moved another step closer to what Martin Luther King, Jr., called the "solid rock of human dignity."

Because of Ruby Bridges and the Little Rock Nine before her and so many others, the color barrier in education was breached, if not completely broken. It would be years before many school districts truly desegregated, but it was finally clear that the United States would never be the same again.

No one knows if the little girl that Norman Rockwell painted was actually Ruby Bridges. No notes were left by the artist, but it doesn't really matter. The painting became a symbol of the civil rights movement's struggle for equality and justice, opportunity and respect—fundamental rights that the rest of the nation already enjoyed when Norman Rockwell put his brush to canvas and sent a powerful political message to millions of Americans.

Ruby Bridges was 6 years old when she walked through the ugly crowds outside William Frantz Elementary. Four years later, I was 7 when Howard Etheridge and I hopped on our bikes one warm September morning and raced around the lake road for the first day of school. For us, it was a big adventure; but for Eufaula, it was the beginning of the end of an era. Howard and I were the first black children to attend Eufaula's all-white Jefferson Davis Elementary.

But unlike Ruby Bridges, we didn't face threatening mobs or an empty school. Quite the contrary. We managed to integrate the second and fourth grades (Howard was two years older than me) without its becoming national news or even local news. I've often wondered how that happened. Eufaula was clearly a town that had divided itself by race, right down the railroad tracks. My uncle Lois, Daddy's younger brother, remembers the last black man lynched in Eufaula. His name was Crockett, and he was hung on a tree in a park near the courthouse jail.

My parents were born and raised in the heyday of Jim Crow, and my hometown wasn't immune to the bigotry this system of second-class citizenship promoted. They didn't call parts of Oklahoma "Little Dixie" for nothing, and the Tulsa riot, in which hundreds of innocent black men and women were killed in a murderous white rampage, remains one of the darkest chapters in our nation's history. So it was neither surprising nor a coincidence that the white elementary school was named after the president of the Confederacy.

Uncle Lois will tell you there were two water fountains on the street in Eufaula, one marked "Whites Only" and one for "Coloreds." He remembers seeing a white fellow drink out of the white fountain and then pick up his dog and let it drink out of the colored fountain. Times were changing by the time I was a boy, but I still remember trudging up the stairs with my buddies to find a seat in the balcony of the Eufaula Movie Theater, the town's only movie house. Blacks weren't allowed to sit with white moviegoers on the first floor, and the local swimming pool was off-limits to blacks entirely. I am grateful I was too young to have been denied service at a restaurant or prohibited from registering to vote or beaten in a civil rights march. Those experiences belong to brave men and women of another generation.

Sometimes, even death can't protect us from the sting of racism.

My wife Frankie's mother had a brother by the name of Nathaniel Jones, but everyone called him Uncle Son. When Uncle Son was just a teenager, his father died suddenly one bitterly cold winter's day while he was working at the local army depot. The body was taken to the local funeral home, and the next day the funeral director told Uncle Son that the ground was frozen, making it impossible to dig a grave.

"We can't bury your daddy, and we can't keep him. You'll have to get that boy outta here now," the director said in no uncertain terms.

Nathaniel was stunned. His father had just died, and rather than comfort another human being in his grief, this man curtly ordered the body out of his establishment. If Uncle Son's father had been white, the family's feelings would have been acknowledged: their father would have been allowed to remain at the funeral home until ground could be broken. Instead, Uncle Son was forced to haul his father's body to the local black church, where it lay for a number of days before a grave could be dug by his family.

Almost every black family in America has a story like this—a relative unfairly charged or physically abused or denied basic human rights. This oral history keeps the black community vigilant against anyone or anything that threatens the enormous economic and social progress it has made despite the kind of pervasive racism that most blacks of my parents' and grandparents' generations experienced. For those of us too young to have been involved in the early years of the civil rights movement, it gives our efforts to create a color-blind society of equal opportunity needed context and perspective. These stories inspire us to finish the job begun by Dr. King and Thurgood Marshall and so many others but also help us understand and take pride in just how far we have come.

To listen to some in the black community today, however, you'd think we were barely out of slavery. We hear plenty of talk about church burnings, the pervasive racism holding African Americans back, and the need for reparations, but the truth is, we've come farther, faster than almost anyone could have imagined 50 years ago when the civil rights movement began.

Fortunately, for my generation, there were some people coming up in the power structure who wanted to see the last of Jim Crow and were willing to fight for that dream. My daddy was one of them, and so was my uncle Wade. They both played the hands life dealt them with honor and integrity. That didn't mean they ever accepted a racist America as a permanent condition. Instead, they fought it on their own terms. Both these men, who began life with little more than a set of solid values, understood that if you wait for the world to change, you'll never change the world.

Sometimes, if you're lucky, an extraordinary individual crosses your path—a larger-than-life figure who changes your life. Dr. King was one of those people for me. My father was another, and so was Uncle Wade. Wade Watts was our family's greatest storyteller, the "keeper of the flame," and one of my personal heroes. Along with my father, he was also one of the most important role models for me as a child growing up, particularly in high school, when I began to develop a deeper understanding of racial issues.

Anybody who ever met them knew immediately that Daddy and Uncle Wade were brothers. Mittie's Indian heritage was carved in Wade's chiseled face and high cheekbones. He was built like Buddy, too—a big, burly man with an even bigger voice—a talent he put to use on a regular basis. Uncle Wade lived in McAlester, about 30 miles south of Eufaula.

Like Daddy, he was a hell-raiser as a boy, and later as an adult, he often strayed from the straight and narrow. He used to talk about his early days selling insurance, cheating either the company or the customer to get ahead. Down deep, Wade knew what he was doing was wrong, and the Lord caught up with him. Like Paul on the road to Damascus, he had an experience that brought him to Christ. He quit the insurance business and eventually became a pastor and a great orator. He turned his life around after coming to the realization that you are defined by the character you exercise and the choices you make in life.

Daddy used to tell us, " If you keep walkin' down a bear trail, eventually you're gonna run into a bear." He was right. I managed to stumble into a bear or two myself growing up, and Uncle Wade was one of the people there to pick me up.

Wade was an activist in the Democrat Party nearly his whole life, but it was the civil rights movement that drove him. As the statewide head of the NAACP, he put thousands of miles on his beat-up old Volkswagen bug, traveling in the bitter cold of our Oklahoma winters and the blistering heat of our Sooner summers. I remember him dropping by Eufaula to see Mittie or Daddy on his way to investigate some civil rights issue. Wade always regretted that he hadn't become a lawyer, and I'm convinced that if he had had the opportunities of my generation, he could have been another Thurgood Marshall. He was that talented.

My first understanding of the "movement" came from Uncle Wade. He spent some time with Dr. King and was so well known in Oklahoma for his NAACP activities that the Ku Klux Klan actually put him on a hit list for a while. I guess the Lord was watching over him, because he never was in the wrong place at the wrong time and lived to a ripe old age.

He and Daddy were always trying to find ways to right the wrongs of segregation in Eufaula and McAlester. The pair of them had been raising Cain for some time with city officials about the exclusion of blacks from the local community swimming pool. There wasn't even a "separate but equal" facility, just a tax-supported resource that kept out a significant number of tax-paying citizens. So, one day, they decided to mount a challenge. From what I understand, Daddy was a determined man that day. I suspect he even had his pistol in his pocket. Growing up, I saw him take it out of the glove compartment of his truck and put it in his pocket on more than one occasion when he expected trouble.

Most of those who were there for the swimming pool showdown have passed away or moved on, so no one is quite sure what Daddy and Uncle Wade did to persuade the powers-that-be in Eufaula to change their ways. What we do know is that 24 hours later, black kids could swim in the pool. My brother-in-law Denzel and some of his buddies were some of the first in the water, and the integration of that pool was never an issue again.

For the kids in the neighborhood, Wade and Buddy's success at integrating the swimming pool was a major event. So was the day we could finally sit in any seat we wanted—balcony or ground floor—in the movie theater. But I was in high school before that happened.

Many influences and pressures forced the coming together of Eufaula's "separate but equal" worlds, but no one event had more effect than the integration of Eufaula's schools. This was the most important cultural change in my hometown during my childhood. In the early 1960s before I arrived at Jefferson Davis Elementary, the equality-in-education earthquake was just beginning to rumble through little towns like Eufaula. The black school, Booker T. Washington, just two blocks from my house, was a plain redbrick building with a series of rooms that had been added on every now and again to accommodate a student body that went from first grade through high school. Cutting-edge architecture wasn't Booker T.'s strong suit, any more than college prep courses were.

All of the town's African American children went to Booker T. trying to get an education with hand-me-down books that came from the white kids' schools when they were done with them. Lack of space, money, and, I suspect, teachers forced the school to combine grades. There was no kindergarten or Head Start in those days, although we certainly would have qualified and could have used the help. Instead, we jumped right into first grade, which was paired up with the second grade. The whole school was structured like this: third and fourth in the same room, fifth and sixth, all the way through high school.

When I began school, nine years after Thurgood Marshall won his great victory for human rights and equal opportunity in *Brown vs. the Board of Education of Topeka, Kansas,* not much had changed in my world in terms of segregated schools. While the community was holding well to the "separate" concept, "equal" was just getting lip service. Of course, Booker T. wasn't my only segregated school; that miserable winter we spent in Wichita, all of us attended a blacks-only school built for the children of the nearby housing projects. But Booker T. is the school I remember most clearly as a young child.

My wife, Frankie, was among the second-graders; and while I had no idea then that we would end up married 12 years later, our time at Booker T. is an experience that shaped both our lives and our views on the importance of education. Like all the teachers, ours was a black woman by the name of Ora Williams, who did the best she could under less than ideal circumstances.

We may not have had the best building or books in town, but Booker T. did top the white schools in one way—its marching band. Without a football field to practice on, the band had to march right through the neighborhood, and we loved it. I'd wait with both arms wrapped around a street sign, listening to the big sound of the instruments trumpeting through the air. With music in my veins, nobody was prouder of that band than J. C. Watts, Jr. Unfortunately, there were no openings for Smokey Robinson sound-alikes, so I was always on the sidelines when the band was marching.

Leaving Booker T. for Jefferson Davis was Daddy and Mama's idea. Like all good and caring parents, they wanted the best for their children, but that meant testing the new integration laws. I've been grateful to them for having the courage to make it happen, knowing that it would go against the grain of some folks in the community.

Brown vs. the Board of Education was a monumental decision that forever changed the futures of both black and white children in the United States. Without that ruling, my children would probably still be learning in separate but unequal schools. But this historic decision didn't integrate the nation's schools overnight, any more than the Emancipation Proclamation freed every slave on the day it was signed. It takes time to change hearts along with the law.

Frederick Douglass, speaking during Reconstruction, said, "The arm of the Federal government is long, but it is far too short to protect the rights of individuals in the interior of distant States. They must have the power to protect themselves, or they will go unprotected, in spite of all the laws."

Supreme Court or no Supreme Court, changing Eufaula from a segregated southern town to a color-blind community was slow-going in the early 1960s. The federal government's reach hadn't extended to the east side of the tracks in Eufaula. My older brothers and sisters were all attending the still-segregated Booker T., but in the fall of 1964, for the first time, black children were given the option to attend Jefferson Davis or Dixie Elementary.

Most chose to stay at Booker T., but my father had had enough when it came to me. He raised eyebrows and blood pressures when he sent me off to Jefferson Davis to begin second grade. It was his way of making a statement: "I'm not going to leave my son in a school where the hand-me-downs are."

For me, changing schools was far less complicated. From my point of view, when I walked into Jefferson Davis, I wasn't making history— just a new set of school friends to go with my black buddies over at Booker T. Most people find this difficult to understand. "It must have been horrible to have been the first black children in a segregated school," they say to me.

As a kid, I don't think the seriousness of the situation fully registered with me. Daddy and Mama probably had a much tougher time that first day than I ever did. I just didn't worry about what the other white kids or their parents thought about the two black faces in the school pictures that year. My teacher was Mrs. Ogden, one of the nicest I ever had. She created an environment in which I was treated just like the rest of the class. There was one moment every day, however, when the separate worlds of Booker T. and Jefferson Davis intersected. At lunchtime, the black children from the school around the lake road were bused over to Jefferson Davis for lunch; and, in a strange twist, I would watch as my black friends headed on into the cafeteria after the white kids, Howard, and I had finished our lunch.

I suppose my only real memory of feeling different was in the matter of my less-than-new clothes. I was a very physical kid (shades of things to come) and also somewhat big for my age. In grade school, I was always splitting a seam or ripping out the knee of my "husky-size" jeans. Many of my white friends would come to school in their newer jeans, always starched and looking good even with an iron-on patch to cover a hole in the knee. Mama and Mittie fixed my jeans in a more old-fashioned way, hand-stitching a piece of denim trimmed from an old pair of jeans. It took longer, but it meant we weren't spending anything extra at the five-and-dime.

There's never any shame in patched clothes, but there is an important message in saving money when you need to and in understanding that material things matter very little. Still, a good psychoanalyst could probably connect my one and only splurge at Chenault's a few years later to the embarrassment I sometimes felt in my old and patched jeans.

I was glad to have the company of kids whose clothes were a little more like mine the next year when several more students transferred from Booker T. By the third year, all of Eufaula's elementary school children were learning to read and write in a totally integrated Jefferson Davis. Booker T. was closed for good in my fourth-grade year. While integration in my hometown was not without its opponents, Eufaula stood up well to what was a difficult task for any community that had existed so long under the yoke of prejudice. That doesn't mean the transformation was totally painless.

The writer James Baldwin said, "Children have never been good at listening to their elders, but they have never failed to imitate them." Every now and again, we'd have a commotion on the playground when somebody would call me a name, like kids do.

When I was 10 years old, I got into a slight altercation with a white kid during a Summer League baseball game. I was pitching when he hit an infield pop-up that arched over toward the first-base line. I tore after it, my legs flying to make the catch. Just as the ball was about to fall neatly into my glove, the batter raced by me and threw me a sneaky stomach punch to make sure I missed the ball. The ump supposedly missed the interference, but the out stood. I was fuming and had no intention of letting what I saw as outrageous behavior go by. So I waited around for the offending player after the game along with four or five of my black friends.

As the white kid headed for home, my posse and I tagged along behind him, hoping he'd turn around and start something. But our glorious revenge was thwarted by the appearance of one of the public high school teachers, who rolled up behind us in her car. She'd already been alerted by her son, who played catcher, to the possibility of a fight brewing.

"You boys need to go on about your business," she said. "Get on away from here."

"You know what he did to me," I angrily answered back. At that age, there's nothing worse than losing face in front of your pals, and I decided to make a stand.

Words flew between the unsympathetic teacher and me. Finally, in frustration, I shot back with, "You think because we're black you can treat us like dogs!"

"Well, you are nothing but a bunch of dogs, a bunch of little puppy dogs," she said, and then she began to bark at us. I was fuming as she drove off. We left for home, and the situation might have ended there, except that her husband told my daddy about it. I was toast.

He jumped all over me. I tried to argue that her name-calling was a racial thing, knowing that would hit a chord with Buddy. "She was calling us dogs," I bellowed. I could see the anger on his face, but he never lost control, not even for a moment.

"One," he said, "you don't talk back to any adults. I don't care if they're black or white. Two, any kind of problem like that, I'm your father, and you let me deal with it."

My parents taught me to turn the other cheek. To walk away from those whose bigotry makes them blind. But I didn't always take their advice. As a boy, especially as a teenager, I took racial epithets as "fightin' words," but I have to admit, it didn't take a racial slur to get me to throw a punch. In my neighborhood, we fought for sport. It was the way we proved our manhood, and while it wasn't the Marquis of Queensberry rules, we had our own code that kept us in check. No one was ever shamed by losing a fight, only by refusing one. What we fought over was as erratic as our punches. For example, we didn't go after each other for taking a verbal shot at each other's fathers. That was just man talk. But say one word against someone's mother, and you better be ready to go the distance.

Sometimes, however, we'd fight just for the heck of it. One Saturday night, some of my buddies and I were out on the street hanging with several older guys. One of them, Eugene, had gotten himself into some trouble and eventually did some time in prison. It wasn't long before an argument broke out as to whether Eugene could "handle" one of our buddies (I'll call him Robert to protect the innocent).

"You can't whip ass on Robert," one of the guys told Eugene. "He's a bad brother, man. Robert will kick your butt!"

That's all Eugene needed. The hunt for Robert was on. Finally, we found him sitting peacefully on another street corner. The poor guy had no clue what was coming. Before he had a chance to even get up off the sidewalk, Eugene had knocked him sideways. The fight was over before it began.

Boxing Golden Gloves saved me from a life of street fights a little later, but as young kids, we thought fighting was a matter of honor. As my own self-confidence grew, however, I came to believe if someone has a problem with the color of my skin, that's their problem, not mine. And someone else's bigotry is never an excuse for failure. After all the silly fights I engaged in to protect my honor, I also concluded that winning a fight didn't make me any more of a man than losing made me less of one.

When I was a boy, racism was a reality in Eufaula. But it would be wrong to think the issue of race dominated my childhood. My life was certainly impacted by race in many ways, but it was never defined by it. Thanks to my parents and the values they instilled within me, my approach to bigotry was, I believe, far more balanced than that of so many young African Americans today.

For me, the question of race boiled down to this: Am I going to let the prejudice of others affect me and what I want to achieve in my life? The answer was no, absolutely not. But there were other questions that faced us growing up in Eufaula in the 1960s. Whenever kids in my neighborhood talked about what they wanted out of life, it was "finish high school, get a job, and get you a car."

I was no different. My parents' lack of formal education limited their own dreams and, in some ways, their hopes for their children as well. They insisted that we finish high school and behave ourselves in the process. But college was just something out there in the wild blue yonder for a long time. As a kid, I thought college was nothing more than a series of great football games that we watched on Saturday afternoons, and the notion that college also meant academic achievement didn't enter my frame of reference.

School was something we did while waiting for summer to roll around. We had a great routine. From grade school through high school, it was studies in the winters. Then swimming, Little League, and chasing down pecan-picking, hay-hauling, or yard work to make spending money through the summers.

That was my "zone," and if the idea of achieving something beyond a high school diploma wasn't on our radar screen, it was because my friends and I lived in a safe, loving, but not very challenging cocoon.

My wife and I have often talked about this. Our own children have always had white schoolteachers, and most of the time they were either the only black kids or one of two or three in their classes until high school. So when they look around, they see almost no one who looks like them. When kids do or say hateful things—and all kids do at one time or another—the sting of being different is even sharper.

At Jefferson Davis and later in junior high and high school, our problem wasn't white classmates, but that most black kids didn't have the kind of black role models in schools to help encourage them to strive to be more than what their environment would normally dictate. I was fortunate to have parents who believed in my ability to make a better life for myself, but that lack of black adults in our school system was limiting in and of itself.

I still love going back to Eufaula and just driving around the streets.

With all its limitations, I appreciate everything Eufaula did for me. I probably missed a lot of opportunities and exposure to the finer things by not growing up in a big city, but I also missed a lot of the turbulence and temptations that urban life can impose on children.

When my pack tested the limits of lawful authority, it was usually with beer, liquor, and maybe cigarettes. We weren't exposed to heavy drugs, and I thank the Lord for that every day. I've never smoked, and I've never had an illegal drug in my system. But many people I knew as youngsters later became alcoholics or strung out on drugs as opportunities dried up and their dreams disappeared. In that sense, Eufaula was like the rest of the country.

When Daddy was a policeman, I once overheard him talking to someone on the phone about a boy he'd caught smoking grass. I thought he was crazy. "Nobody is dumb enough to take grass or hay and smoke it," I thought. I didn't know what a mixed drink was until I got to college and saw people ordering piña coladas and daiquiris. In Eufaula, if you couldn't spell it, you didn't drink it. We could spell beer, even though we couldn't get hold of it very often. Whenever Mama suspected I might be sneaking off to share a six-pack with a couple of buddies, she'd yank my chain and tell me, "Boy, I'm not going to tolerate you trying to be mannish." She really meant I was just a little boy trying to act like a man. But all of us kids were guilty of that from time to time.

At that stage in my life, our most serious crime usually revolved around stealing fruit from some unsuspecting neighbor down the block. There were some great apples, peaches, and cherries to be had if you could run fast enough. Maybe that's where I got my speed as an athlete, staying one step ahead of Charlie Washington, the owner of the best plum trees in the county. One day, when I was in fourth grade, my buddy Paul Ray, his brother, Steve, Charles Perkins, and I were busy stealing a few ripe plums, when Mr. Charlie bolted out his back door waving a shotgun and threatening us with "serious trouble" if he caught us there again.

"Yes sir, Mr. Charlie. We're sorry," we told him looking appropriately repentant. But a little buckshot wasn't going to keep us from Charlie's plums.

I had a plan.

"Look," I told my friends, "When it rains, Mr. Charlie won't come out of his house. That's when we strike."

So the very next time it poured rain, we swooped down on that unprotected orchard and made off with all the plums we could eat, leaving a few to rot on the ground out of pure spite. Our alternative source of fresh fruit was the backyard of a lady everybody called "Ms. Charlie." She had a mean disposition, especially when it came to her apple trees, which were the finest around. We learned that it was generally best to call on her orchard after 11 at night.

Our days consisted of swimming or playing football on the church grounds or shooting hoops when we weren't speeding around town on our bicycles like a band of hooligans, which I suppose we were in a benign sort of way. No matter how much we promised our parents to stay out of trouble, trouble often found us. A lot of our time was spent down by the lake, where we'd bust up beer bottles. When we were feeling particularly daring, we'd steal a snack from Hammonds' Grocery, a neighborhood store.

In seventh grade, I learned how to shoot dice under the streetlight at Hammonds'. There was a huge rock right next to the store, and you could find a crowd of kids gathered there with or without a pair of dice most evenings. It was while working at JM's that the finer points of gambling were explained to me. By ninth grade, I thought I was ready for Vegas, but my luck was about to run out. It was a summer evening, and I'd headed down to the rock after a long day of doing yard work. I was tired but feeling good about the $30 I'd pocketed at quitting time. Lady luck deserted me that night, however, and I managed to lose $21 of that hard-earned cash. I also lost my interest in gambling, and I haven't picked up a pair of dice since.

Probably the dumbest thing I did as a kid was my one joyride in a "stolen" car. Pee Wee Harris, who lived on our side of town, was a man with two weaknesses—drink and the lady who owned the corner store. He'd pull up to Hammonds' to court his lady friend, leave the keys inside his old pickup truck, and head into the store for a spell to eat and then sleep off his buzz.

Hanging out behind the big rock near the store, we soon were on to Pee Wee's nocturnal habits and that got us to thinking. Once he was in the front door, we knew Pee Wee was down for the count. Nobody would see his smiling face for hours. Meanwhile, that old truck just sat there—one big temptation for a bunch of kids with time on their hands. One night, four of us abandoned every ounce of common sense and headed straight for the truck. We were eighth-graders, and I was the biggest, so I was elected to get behind the wheel.

Even in his drunken stupor, I'm still amazed that Pee Wee didn't hear the gears grinding as we took off for parts unknown. It was a miracle that we didn't end up in a ditch or worse. Amazingly, it never occurred to us "four musketeers"—Paul Ray, Popcorn, Tony Smith, or me—that our little jaunt could get us jail time or that anyone who saw a pickup with four junior-sized knuckleheads in the front seat screeching their way down the road might decide to alert the police. And there was one policeman in particular I sure didn't want in my face.

Actually, if either of my parents had found out, I wouldn't have been able to sit down for a month. We were lucky that night and got the truck back with no one the wiser—including us. We loved to brag about that joyride, but the truth is, we were scared silly the whole time and never did it again. By the way, this was how I learned to drive a stick shift. It gives new meaning to the phrase "school of hard knocks."

With Buddy and Helen Watts looking over your shoulder, you didn't get away with much, believe me. And they weren't alone. There were plenty of parents just like them in east Eufaula. It was a given that we were supposed to behave ourselves in school and elsewhere if our parents weren't around. That didn't mean we always did.

Hillary Clinton and I share very little when it comes to philosophy and values, but I wasn't one of those who criticized the title of her book, *It Takes a Village to Raise a Child*. In fact, for all practical purposes, Eufaula was and still is a large village, and folks look out for each other and especially for each other's children. From my perspective today as a parent, I don't think there's anything wrong with that notion, although I had a slightly different view when I was the regular subject of neighborly surveillance. Anything we did within a 5-mile radius of home was very likely reported back to Mama and Daddy before we'd even had time to cook up a remotely credible explanation. Sometimes, we caught a tongue-lashing right on the spot from one of the neighbors, but we always knew it would be worse when we got home. It usually was.

The Bible says friends are a "strong defense and a treasure." That's certainly how I felt about mine. The crowd I ran with included Billy Mac, who had a sense of humor like Richard Pryor's, always messing with folks, laying them low with some quick, slanted observation about how they looked or how they talked. Billy could have been one heck of a boxer or a good college football player, but he got homesick after one year of college and dropped out.

Then there was Jerry Shine, probably the smartest and most industrious of the bunch. He was very quiet, but he earned himself a special place in our eyes because he was the first to get a car that he paid for all by himself.

Jerry Jerome, a cousin of mine, was the loudmouth of the group. He was potentially the best tight end ever to play at Eufaula, but Jerry didn't want to do things the way the coach wanted them done, and he didn't last. Then there was Mike Jones. We called him Popcorn because that was his favorite food. We also called him "Punybird" because it seemed to us like he boxed in the 65-pound weight class from the time he turned 8 until he reached 15. We kept telling him that popcorn wasn't the best way to put on weight.

Jim Dandy was Jerry's little brother. When he and I were in high school, we both worked evenings at the Arrowhead Lodge on Lake Eufaula. We started right after football practice and stayed until 11 most nights, which was late for a pair of guys in training. Even worse, we sometimes had homework to do when we dragged our weary bones through our front doors. One night, we left 20 minutes early and asked Gary McNeal to punch our time cards for us. Naturally, we got caught. The next day, the boss called us into his office.

"You're both terminated," he told us.

I didn't want to admit that I didn't know what the word meant. With my usual bravado, I assumed this was nothing more than a temporary setback.

Later, when Jim Dandy told his mother we'd gotten fired, I corrected him.

"No, we didn't," I told her. "We just got terminated."

"That means fired," she said, rolling her eyes and shaking her head.

"Oh." Education comes from unlikely sources at times.

There was Paul Ray, who was a pretty good hand at boxing, and also my pecan-picking buddy. Oscar B. and I had paper routes together. His real talent, however, was his mechanical ability. Oscar B. could do more with a broken bike than any kid in the neighborhood. With a few bucks' worth of parts and some odds and ends salvaged from the trash, he could put together a perfectly good bike that would last him a long time—all without a government research and development grant.

Gary McNeal was a heck of a basketball player, but he wore a big Afro and didn't want to cut it. Eventually, he quit the team because of the rules concerning long hair. I wore my hair the same way for a number of years, too, and Daddy hated it. Like most teenagers always fomenting a little rebellion, Buddy's disapproval of my hairstyle was all the excuse I needed to keep me away from the barber.

Every time I heard Daddy say, "You need a haircut!" I ran the other direction. When I did finally break down and get it cut, usually on coach's orders, Gary was the man to see. Ironically, the boy who wouldn't cut his own hair was a master barber. Working with clippers and a handheld razor blade for edge-ups, he could make any one of us look pretty good. To his credit, he stuck by his guns on the hair issue. It's even longer today, although a ponytail has replaced the Afro. Ironically, he works at the sister lodge of the Arrowhead where I was "terminated," but he's got a lot better track record staying employed there than I did.

Big Howard Etheridge, my partner in integrating Jefferson Davis, was one of the purest basketball shooters to play in Eufaula in my time. If we could keep Howard eligible, no easy task, and he got hot, he'd hit nothing but net. It didn't matter how many defenders they threw at him, Howard was unstoppable on a streak. In those days, he was our main hope for someone from the 'hood playing college basketball. A scholarship took him to Clarendon Junior College in Texas, but after one year, he decided to call it quits.

Some of my friends have done well with their lives. Others haven't been so fortunate. In a town where both educational and job opportunities were limited, far too many of the people in my neighborhood were left behind. These were boys and girls with enormous potential but without the means to reach it. When I was growing up, most black kids in Eufaula didn't go to college. Some made it and did well. Others won athletic scholarships and lasted barely a year.

The roadblocks that faced my parents and me, to a lesser degree, were far different from those today. Given the times in which they lived, my father and uncle were both considered successful. For a man with only 6 years of elementary education, my father worked wonders with his life. He even won a seat on the city council a few years after I was out of school. Uncle Wade's life was a testament to the power of one individual to change the world around him.

Despite beginning my life in segregated schools and in a community that still separated black from white, I was blessed. I was able to rise above poverty and prejudice to achieve my goals. As I've said, I have my family to thank for that, along with my education and some great teachers, coaches, and friends. Mama and Daddy gave me the right values. They believed in the ability of each of us to determine our own destinies. They believed in the power of education and put me in the best school possible. And thanks to those who bravely fought for civil rights, the obstacles that prevented their generation from being all it could be no longer stand in the way of my children.

Yet we're seeing a disturbing trend in the black community today regarding education. No one could argue that educational opportunities for African American children have improved dramatically across the country. We still have more to do, particularly in economically depressed areas. Nevertheless, the progress is remarkable.

We know education is still the path out of poverty, but in an ironic twist, as opportunities have increased over the past quarter century, so has the disconnect between many black children and learning. I could never claim to have been a Rhodes scholar, but I understood the value of education from my parents. Schoolwork got done in my home, and done on time. Teachers were respected and so was the school environment.

One of the most disheartening aspects of African American life today is a cultural phenomenon in which doing well in school is seen as being "white." School isn't cool in the eyes of far too many black students. Where this trend came from, and how it came about, is debatable, but the results of this self-defeating attitude are plain for all to see. African American children are underperforming, and in doing so, they are putting limits on their ability to succeed.

This trend, unfortunately, is seen in black students from both urban and rural settings, from poor *and* well-to-do families. John McWhorter calls it an "anti-intellectual current in black American culture" so strong that "the black person who chooses to truly embrace school has indeed had to all but leave the culture." I am becoming increasingly convinced that this anti-learning attitude is a direct result of the "blame game" played by many traditional black leaders today. The problem with the academic performance of black students is poor schools, they say, or racial bias in testing, or welfare reform. Second-rate schools do exacerbate the problems facing black students, but there are plenty of examples of kids from low-income families and schools in depressed areas who have succeeded in spite of financial challenges. And we are also seeing lower test scores from black children in upper-income neighborhoods, where the schools are first-rate.

Race may still impact black children today, but black children don't have to let race define them or their dreams any more than it did mine. The doors that were closed to their grandmothers and grandfathers are wide open for this generation, but no one can take the first step for them, and only victims wait to be carried over the threshold. That's the lesson they should all learn from Ruby Bridges and Wade Watts and from those who marched in Selma and walked with Dr. King in Washington.

My childhood wasn't perfect, but if I had waited for life to be perfect, I would not have become a member of the U.S. Congress. I loved growing up in Eufaula. I never saw myself as the "victim" of an economically depressed area. The times I spent hanging out with my buddies were some of the best years of my life. Booker T. Washington and Jefferson Davis Elementary bring back only warm memories of the joy of learning from dedicated and kind teachers.

Eufaula was a wonderful place in so many ways, but as I grew, somewhere inside me I began to hear another voice, and nothing in my raising taught me that I couldn't follow my heart wherever it led. I always knew I wanted something more than my hometown had to offer. But what was I looking for? How did I answer that voice? In the next few years of my life, I was to discover the talents that would take me to a world I didn't even know existed.

6

Leaders May Find Mountains to Scale, but Only Teams Can Truly Move Them

> Sports is the ultimate passport—it transcends all races and creeds, even nationalities.
>
> Lesley Visser

Apart from the manmade lake on the eastern edge of town, about the biggest thing that Eufaula ever produced was a trio of massive and highly athletic brothers, Lucious, Lee Roy, and Dewey Selmon. During my elementary and junior high school years, the Selmon brothers made Eufaula High School's Ironheads one of Oklahoma's most rugged football teams. They were the closest thing to genuine heroes I knew.

One by one, the Selmons earned athletic scholarships to the University of Oklahoma, where they anchored the most fearsome defensive line in the nation. By then, Lucious, the oldest, stood about 5 feet 11 inches and weighed 240 pounds—he was the shrimp of the family. Dewey grew to 6 feet 2 inches and weighed 250 pounds, and baby brother Lee Roy, the gentlest of the trio, topped out at 6 feet 3 inches and 260 pounds.

I heard a wonderful story about Lucious when I followed him to OU on a football scholarship. It seems one of the athletic department secretaries, Cecilia McEuen, had a couple of boys in grade school. One day the youngest got into a schoolyard fight, but unfortunately, this little guy didn't have Gwen Watts on his side. His older brother refused to come to his rescue, believing he needed to learn to fight his own battles. All the women in the office argued that the older brother should have protected the smaller boy; all the men said he handled it right.

Just as the discussion started to heat up, Lucious Selmon walked in. With two younger brothers, Lucious seemed like the perfect choice to end the debate.

"Lucious," Cecilia asked him, "what would you do if somebody was beating up your younger brother?"

He thought for a moment, then smiled and said, "If somebody was beating up *my* little brother, I'd *run!*" Everyone loves that story almost as much as Lucious. He was not only one of the greatest football players Eufaula High ever produced; more important, he is one of the nicest guys you'd ever want to meet.

I was lucky enough to know all three of these amazingly talented football-playing brothers. They were great friends who came from a background much like mine. The Selmons didn't have much money, but everyone in Eufaula knew them as one of the kindest and most hardworking families around. Like so many, they made their living off the farm, but in their case, they did it without the benefit of a tractor. Hand plowing, sowing, and weeding is tough, hard work, but with nine children, there was never a labor shortage at the Selmon place.

I don't know whether it was all that hard work or something Mrs. Selmon put in *her* beans, but nobody produced world-class football players like the Selmons. By the time their playing days were over, collectively, the three brothers had won some of college football's highest awards, including the Lombardi Trophy and the Outland Trophy. All three were All-Americans. All three played pro ball, and Lucious ended up as an assistant coach at OU, a job that I had once thought might be a part of my future.

As children, we never know what will push us in a particular direction. Frankie says I used to talk about being a dentist. I remember talking about being a lawyer. I know I dreamed about being a singer, but it was Lucious Selmon who opened my eyes to a whole new world of possibilities. I remember the moment as if it were yesterday.

It was Thanksgiving Day, 1971, and the national college football championship was up for grabs. Nebraska was ranked number 1 in the nation. Oklahoma was number 2—at least outside the borders of the Sooner state. These two powerhouse teams were set to face each other in what was being called "The Game of the Decade." That Thanksgiving you would have been hard-pressed to find a football fan anywhere in the country watching anything but the Nebraska-Oklahoma game, and I was no different. Sprawled on the living room floor in front of the TV, I had a front-row seat for the action. The old green couch where Lucious Selmon had spooned with my sister Gwen a few years earlier was still in residence, with Buddy stretched out after getting his fill of turkey and mashed potatoes. Of course, Daddy ended up sawing logs long before the game was over. Football didn't mean that much to him then, even though both Melvin and Lawrence were terrific players in their own right, with college potential.

But I couldn't take my eyes off the set. It was as if a curtain had risen and an entirely new world was opening before me. There was Lucious Selmon rolling onto the field with "Selmon 98" stretched broadly across his back in tall letters. He was just a sophomore then, playing noseguard instead of his Eufaula Ironheads position of fullback, and he looked like a conquering hero to me.

To see someone from home come thundering into a stadium on national television with thousands of people wildly cheering him on—well, that was nothing less than a life-changing event for this 13-year-old boy. Every time Lucious crossed the TV screen, making a tackle or just coming off the sidelines, my adrenaline pumped like I was right there with him. I wanted to memorize every second of the game.

The following week, nobody in Eufaula talked about anything else but the big game and our hometown hero. Lucious Selmon had become community property.

"Did you see Lucious on TV last Thursday?" was all you heard on the playgrounds and at church and in the local watering holes. I heard it over and over again at JM's as I cleared tables. I had to laugh, though, at the number of customers who "knew" Lucious after that game. He probably made more personal acquaintances that week than in his first 19 years combined.

I will never forget that Thanksgiving Day. Nobody I knew personally had ever been on TV. When I saw Lucious, it completely changed the direction of my life, and my hopes for the future. I realized I could set my sights higher than just getting a job and my own car. I began to think, "Maybe that could happen for me—go get an education and play big-time football," and the self-imposed limits of living in Eufaula seemed to disappear with the echoes of that Oklahoma-Nebraska crowd. Eventually, the talk in town died down, but not my dream.

Yogi Berra said, "If you don't know where you're going, you'll end up somewhere else." After that game of the decade, I knew where I wanted to go, and luckily, my high school years provided me with a roster of outstanding mentors who taught me to believe in myself and my dream and to embrace the right values for a successful life. Short of my family, no one had more influence on me than Paul Bell, the Eufaula High head football coach, and through him I discovered three very important things.

I came to realize that I had a unique talent—a physical ability that, if respected and nurtured properly, could take me wherever I wanted to go.

I learned the true meaning of leadership and to believe in my own ability, not just as an athlete but also as a potential leader.

And, finally, I came to understand that leaders may find mountains to scale, but only teams can truly move them.

These three lessons formed a powerful epiphany for me and have guided my life ever since, and I have Coach Bell to thank for them.

I was just 4 years old the day Paul Bell came to Eufaula in 1962. He met my daddy when looking for someone to move his family from Stigler, about 30 miles away. When it came to hauling, nobody could beat Buddy Watts, and the two began a friendship that lasted nearly 40 years. Paul had played on the 1958 NAIA National Championship team at Northeastern State in Tahlequah and never lost his love for the game.

At the time he was hired, Eufaula was hungry for the football glory it had enjoyed in the days when "Ironhead" Hanson made our teams feared all across the state. Paul's mandate was to bring back those good old days. He accomplished his mission and then some. Not only did he make Eufaula High one of the high school football powerhouses in Oklahoma; he became a force for integration in Eufaula as well.

When Coach Bell arrived on the scene, the schools were still segregated. It would be two more years before Howard Etheridge and I would enter Jefferson Davis Elementary. In that fall of 1962, Eufaula High had 135 students, but only 18 kids went out for football—all of them white. Paul spent the next summer coaching baseball games and convincing kids who were about to enter Booker T. as freshmen that they would be welcome at Eufaula High.

Even though the first moves toward integration had begun in Eufaula by this time, and despite the fact that educational resources given Booker T. were demonstrably inferior, there was an understandable hesitation on the part of the black kids he was recruiting and their parents. Eufaula High was uncharted territory where African Americans had never been welcome. Coach Bell's enthusiasm overwhelmed their doubts.

His power of persuasion brought blacks and whites together to create a winning team, but he wasn't just a good salesman. With Paul Bell, you sensed he was a stand-up kind of man whose word you could take to the bank.

Today, he'll tell you there was no problem with the kids when it came to integration. "Kids are kids," he says. "You teach segregation. That's not an inherited characteristic—just like love and hate."

The first year, four black kids played for Eufaula High, though most opted to remain at Booker T. for classes. The next year, seven or eight went out. Just like with Jefferson Davis, by year three, the football team was totally integrated. My brothers Lawrence and Melvin were among those who traded Booker T. for Eufaula High, paving the way for me a few years later.

All those summers tossing around 75-pound bales of hay had given my brothers the kind of muscle power that's invaluable on a football field. Coach Bell says Lawrence was one of the best players Eufaula High School ever saw and he could have been NFL material. Like me, Lawrence has a soft spot for Coach Bell. "I liked the coach," he said recently. "I don't remember anyone else standing up for us the way he did."

As for me, I had begun to show some signs of athletic ability by fifth grade. Every day at lunch, we would walk a few blocks to another school's cafeteria. For us kids, the noon-hour walk soon became a daily race, and I won more than my fair share. I also loved to shoot hoops and play baseball as well as our informal version of football, and I even boxed for three years, going 7–0.

Athletics also gave me some leadership roles early. Sometimes, I was the resident cheerleader of the team. When I played second base or catcher, I had the annoying habit—at least to the other team—of yelling, "Try me!" at whoever stepped into the batter's box. Other times, I found I could lead by just going out and getting my job done.

I certainly didn't see myself as some kind of athletic superstar, and neither did my buddies. Eventually, I came to appreciate the talents God gave me, but in those days, baseball or football or basketball with my friends involved more horseplay than playbooks. We were overdue for a little structure, and Coach Bell and his coaching staff were just the folks to provide it. By the time I reached seventh grade, the coach had begun his "farm system," a unique program to groom the future athletes who would soon make up his talent pool in high school. Whatever the motivation behind the program, it worked like a charm. Every kid who wanted to be involved in sports of any kind enrolled in seventh-grade athletics. We were taught the fundamental skills of whatever sport was in season.

Day one consisted of a short course in strength-building exercises in the weight room. Apparently, Coach Bell hadn't heard of Buddy Watts's hay-hauling regimen. During football season, we spent a week just learning to pass the ball, then another on kicking, then on catching and running. We learned the basics for every position in every sport. We were ready to compete across the entire range of school-sponsored sports.

Seventh grade was also the year I saw Lucious Selmon on TV. From that day on, my consuming focus was sports. It wasn't long before I began to get positive feedback from my coaches. They saw the potential in me long before I did, but they also sensed I was the type who could just as easily take a wrong turn and run into that "bear" Daddy used to talk about.

It was no secret I was still out on the streets at night, chasing around with my buddies, generally on the verge of some kind of trouble. Much as I loved sports, I couldn't seem to give up the high life completely.

Finally, Lamar Armstrong, one of the assistants, took me aside and told me, "Watts, you can do the same thing the Selmons have done. You could even make Eufaula forget the Selmons." I knew that he was embellishing a bit for my benefit, but he hit a nerve. Then, he got to the real point.

"But," he told me, "you can't be on the streets all the time. You've got to make good decisions."

To a sports-loving kid, a coach is a high priest, and when he or she talks to you as an individual, not just as part of the squad, it makes you feel like you've arrived. You're not just one of the scrubs anymore; you're somebody they want to be able to count on. Robert Newton, the other Eufaula High School assistant coach, gave me similar advice.

"J. C.," he said one afternoon at the end of a sweaty practice, "you've got to keep your head screwed on right. You've got to have a good attitude." They both knew that Paul Bell as well as Perry Anderson, the varsity basketball coach, would soon be demanding tight teamwork and plenty of application from me. As coaches, they were trying to help me build that foundation and, like my parents, instill in me the life values that would make me a successful athlete and human being.

I started playing organized football in the eighth grade. The coaches wanted me to play quarterback, but I was against it. Lucious had played fullback, and I wanted to do the same. I figured, if it worked for Lucious, it would work for me. In those days, we ran the straight T offense, which was built around the fullback.

By ninth grade, though, I was hanging around after practice throwing the ball, and that caught the coach's eye. Next thing I knew, they had added a passing play for the tailbacks, and I discovered I could whip the ball upfield pretty well.

It was in the tenth grade, however, that my football career changed course in a dramatic way—as dramatic as high school ball gets, anyway. That fall, the team had two quarterbacks. One could run and the other could pass. What the Ironheads needed was someone who could do both. Coach Bell tapped me for the job. It was a controversial move, and the coach took some heat for it. Nobody with my kind of complexion—

and a sophomore to boot—had ever filled the quarterback position at Eufaula High.

I had two days of practice at quarterback before the game with Spiro that week. I was still at fullback when the game began. The coach started one of the regulars, but by the opening minutes of the second quarter, we were losing 22–0. Finally, Coach Bell turned to me with the words I had been expecting to hear. "I want you to go in at quarterback." I didn't know much, but I had at least a basic understanding of how to run the offense. By halftime, I had thrown three touchdown passes, and we had fought our way up to 22–20.

We lost that game, 36–20, but I remained at quarterback for the rest of my high school career. One of the other quarterbacks moved to half-back and the other quit the team. I later heard from some buddies that the guy said he wasn't going to play behind a black quarterback. A couple of other white kids went with him.

Coach Bell never talked to the team about their decision to leave. The coach never spoke about race directly. Instead, he relied on the two straightforward principles that he applied in any and all disputes. "The Ironheads are my team, and I'm going to do what's best for the team."

It was a couple of weeks before someone let me in on the reason for their departure. I'm not sure I really understood the significance of the coach's decision at the time. A black quarterback? What's the big deal? It just seemed logical to me that everyone should be playing the position they play best.

I eventually learned there had been quite a lot of grumbling around town over my promotion, but people protected me from that ugliness at the time it was happening. If Eufaula was home to a few bigots, it was also home to a lot of good people who didn't care what color I was if I could get the ball across the goal line.

Actually, the funniest protest came from Max Silverman, a local businessman and president of the Quarterback Club, an athletic boost-ers club. At first, he had complained to Coach Bell that moving me was a mistake, not because I was black but because he thought the team needed me more at fullback. Once we started winning, however, Max became a convert. He buttonholed Coach Bell in a meeting and told him, "You're still not a very good coach!"

The coach was a bit taken aback. "What do you mean?"

"If you were any damn good," Max cried, "you would have moved Watts sooner!"

Coach Bell told one writer he made the change because he wanted the best football player on the team at quarterback.

"We've got to have a kid that's got a little maturity. We've got to have the best leader," he said. For some in Eufaula, race had been an issue, but it did not define me for most. My talents did.

Having been in the thick of integration in Eufaula, it's not surprising that Coach Bell would be the first to tap an African American for quarterback or that he would expect more of his teams when it came to race relations. There's a story that baseball writer Donald Honig tells about Jackie Robinson's entrance into major league baseball. Branch Rickey, the general manager of the Brooklyn Dodgers, wanted to sign Robinson, so he asked the great outfielder to come to New York for a meeting.

Rickey told Robinson he was looking for more than a great player to break the color barrier in major league baseball.

"I'm looking for a man who will take insults, take abuse, and have the guts not to fight back . . . [because] that would set the cause back twenty years."

What a recruiting pitch! But Robinson knew what he was asking him to do and why. Enduring the racist reactions that were sure to follow would be tough, but being the first African American in major league baseball would certainly change baseball and maybe the nation forever. It ended up doing both.

Robinson thought hard for a few minutes and then told Rickey, "If you want to take this gamble, I promise you there will be no incident." We all know the rest of the story.

Coach Bell taught us to exhibit that same kind of grace and maturity when we faced the occasional bigot on the field. We played a couple of all-white schools and once in a while, one of the opposing players would make a racial remark to one of the black Ironheads.

It would have been understandable if we'd laid into the offender, but Coach Bell's philosophy was a steadying influence on the team. He taught us better than to respond in kind. "Look, they're getting their ass beat and trying to make you feel lousy, too," he'd tell us. "Nobody who'd amount to anything would use a derogatory term toward any human being. Just don't pay attention to them." A great lesson learned that has served me well over the years.

Coach Bell treated me with respect, and I appreciated it. When I was a sophomore, the coach would work with me at the chalkboard teaching me how to call audibles. He taught me other things that went far beyond football, too. He would often say, "The more people think of you, the more humble you should be." He'd tell me, "Don't try to please everybody. Not every soul is going to be your friend, and that's okay." And "Don't be rude to people, and don't worry if not everybody likes you."

Be your own man was good advice to a fledging football player. So was his emphasis on team play. He was clearly grooming me to lead, but he never let me forget that the team was more important than any one individual.

Maybe I was naïve, but the quarterback controversy seemed like the classic tempest in a teapot to me. I think it actually had a positive impact on the players in the end. We realized that as a team, we were stronger than the controversy, and as individuals, we were totally committed to doing everything possible to become better competitors. We blossomed beyond expectations.

Early in the season, we had lost 32–12 to Checotah, which was ranked number two in the state for our division. Later, on a weekend when we played a particularly outstanding game and upset one of the league's most powerful teams, Salisaw bumped off the mighty Checotah Wildcats and moved into the number one ranking in the state. We got the news when we got off the bus, fresh from our victory and feeling pretty good about ourselves. The feeling didn't last long. The Ironheads were scheduled to face Salisaw the following weekend.

Taking on the number one team in the state with a green quarterback might have led most coaches to lower expectations just a little. Not Coach Bell. He said confidently, "Next week, we're gonna beat Salisaw. We're gonna give 'em a knock."

By the time Friday rolled around, he had a team of believers—me included. Every time I'd seen him that week, I'd yelled, "Hey, Coach. We can beat 'em!" Years later, he admitted to me that a voice in his head always answered, "Yeah. And pigs can fly."

He never let us think for one moment that he had any doubts, however, and by halftime, the Ironheads had jumped to a 23–0 lead. At the start of the third quarter, I got loose and ran 76 yards down the field. I can still remember how my legs seemed to lift off the ground that night as I flew past Salisaw's vaunted defense toward the goal line.

At that point, Coach Bell turned to one of his assistants and said, "There's no sense in us calling plays. Let the team call their own!" And we rolled to a 41–7 win.

As the season progressed, my respect for Coach Bell grew, and I found myself hanging on to his every word the way a would-be architect might follow after Frank Lloyd Wright, waiting for the wisdom to flow. That time the coach had spent teaching me audibles turned out to be both a blessing and a curse. I loved sizing up the opposition's defensive schemes when I took the snap from center. Once I'd pegged their weakness, I'd call an audible to maximize our advantage. There was a problem with that strategy, however. The team trained on standard plays, and my variations had a tendency to confuse us as much as the opponents.

Once, after I'd called three or four audibles in a row, and we'd drawn two offside penalties, Coach Bell motioned me over to the sidelines.

"Why're you calling all those audibles?" he asked me.

"Coach, it's open."

"Well, you've got most of the sophomores scratchin' their heads. Every time they stop to think, they can't get themselves going again. And another thing. You see those folks up there in the stands watching me talk to you?"

I looked up at the bleachers. As usual, they were completely full. Football is as popular in Oklahoma as spending money is in Washington. Everybody joined in the fun.

"Yeah, Coach. I see 'em," I said.

"Well, they're gonna figure out that you're calling all the plays and they don't need me to coach. They're gonna fire me if you keep this up." I cracked up.

He could have read me the riot act. I had let my ego get the better of me. Instead of leading the team, I was trying to dictate to it. Coach Bell made me laugh, but he made his point. I eased up on the audibles until the other players caught on a little better. I learned that while the leader may determine the speed of the pack, *he* gets nowhere if the team doesn't know where to go or why.

By the end of the season, we were much more capable of improvising. We had made it as a team, and I had come to understand that the trust of your teammates is a pretty special commodity. You don't waste or abuse it.

For me personally, being the quarterback of the Eufaula football team was a thrill. But I discovered the real challenge and satisfaction was in *leading* a football team, and that knowledge changed my views of both my own abilities and the opportunities open to me.

During my high school career, the Eufaula Ironheads compiled a 25–9 record, and I left Eufaula High proud of my accomplishments on the gridiron: 7,000 total yards, being named a high school All-American, and perhaps most important, voted captain of the team. But I knew that whatever reputation I had achieved for myself as a football talent, I could never have succeeded without the other 10 men on the field.

The movie *Remember the Titans* tells the story of the integration of a Virginia high school through the melding of white and black football players into a championship team. Initially the team doesn't come together and neither group will block or protect the other. The key black player, Julius, and the white captain, Gary, finally come to terms with each other when Julius tells him, "I'm gonna look out for myself and I'm gonna get mine."

"Man, that's the worst attitude I ever heard, " Gary angrily spits back.

Julius looks him in the eye and says, "Attitude reflects leadership, *Captain*." At that moment, Gary understands not only what a team really is, but how to lead it. As captain of the Eufaula Ironheads, that was something I had to learn, too, and did. Whether it's playing on a sports team or being a member of a political party, I learned when you put on a jersey, you act like a team member. You put the interests of the team first. That doesn't always happen in Washington, and it is one of the things I have liked least about politics.

People forget they owe loyalty to the "team." I'm not suggesting blind loyalty, however. I don't always agree with my party on every issue, and none of us should ever forget that we are Americans first and Republicans and Democrats second. But there are ways to live up to your own beliefs and principles, do what's right for the country, and still support the team. I only wish everyone in Washington could have experienced the kind of team ethic I was privileged to know and the coaches who made it happen. The country would be better off for it.

Before Coach Bell came to Eufaula, the Ironheads were at the bottom of the pile. They were losing games 70–0, and spirits were pretty low. Eufaula had been one of the strongest football clubs in the state, but then it hit on hard times. Just a few years after Coach Bell's arrival, however, the Ironheads' Quarterback Club was showing winning game films.

One evening, one of the members, who had had words with Coach Bell when he had first arrived, came up to him and said he had a confession to make.

"I'm not good at this," he told Coach Bell, "so let me talk, and you be quiet." He then went on: "I thought you were a little smart-aleck jerk when you came here, but I want to tell you something, and I ain't gonna say it more than once. I think you're a hell of a football coach, and the superintendent better raise your wages or we're gonna lose you."

After that, he and Coach Bell were friends for life.

I've worked with some great coaches and athletic teachers in my life. Coach Bell was pretty special, and so was Perry Anderson, the varsity basketball coach.

Anderson and Bell were eventually named to Oklahoma's Coaching Hall of Fame, and no one was ever more deserving than that pair. If you drive up to Eufaula High today, you'll be driving on Anderson-Bell Street. As I said earlier, fate dealt me some outstanding mentors.

Coach Anderson stood out for more than just his coaching abilities. He is a Native American, with blood from the Choctaw, Chickasaw, Cherokee, and Creek tribes, and, as a young man, he played basketball at both Oklahoma State and Northeastern State. Coach Anderson had also been an All-State baseball player in high school. For 10 years he played on a number of Yankee farm teams, never making it to the "big show" but never regretting a moment, either.

As outstanding as the football was at Eufaula High School, Coach Anderson shaped up some highly respectable basketball teams, too. At one time, all of the Selmon brothers played on the squad. Driving the lane against these future NFL bruisers was a life-changing experience for any opponents who were crazy enough to try it.

I loved playing basketball almost as much as football and did my best shooting from 20 feet in. My style was to use the glass a lot and follow through. I was helped by the fact that I could shoot left-handed or right-handed. We weren't the biggest team on the courts, but during my three years on Coach Anderson's team, we won our share of games. Our point guard was the coach's son, Greg, who stood 5 feet 7 inches. I played center at the whopping height of 5 feet 11 inches.

Throughout high school, I kept my head crowned with a full-fledged Afro—the one my Daddy hated. Coach Anderson used to razz me about it on a regular basis.

"J. C.," he once laughed, "if you cut that Afro, we might find out that you're really just a midget."

We qualified for the regional playoffs every year. In my final year, we played Seminole, whose team had two guys at 6 feet 6 inches, while their smallest starter was 6 feet 4 inches. We felt like we were playing in the Land of the Giants. It took them three overtime periods to put us away, however. Coach Anderson made us believe that size didn't matter, heart did.

Coach Anderson knew all about an oversized heart. He grew up in a hardworking family and knew how tough it could be to come up with money for extracurricular activities. He established a fund at Eufaula so that *every* boy who wanted to compete in a sport and every girl who wanted to be a cheerleader could do so without worrying about whether their folks could afford their outfits and equipment.

Like Coach Bell, Coach Anderson was instrumental in the integration of Eufaula High. Basketball provided one more bridge between races that brought us together. By 1967, times were changing. For the first time, the boys on the basketball squad nominated and voted for a black girl as basketball queen. Her name was Elaine Basset.

This was another instance where Coach Anderson came to the rescue. Before the vote, the principal at the time asked Coach whether he thought there would be a problem if a black girl ran. Coach was all for it, but when she won, he got a call from her father.

"She can't be the queen because we don't have the money for the gown and all those kinds of things," her father explained.

"Don't worry about the money," Coach said. "She's the queen. She's going to have everything every other queen has always had." And she did.

Along with Coach Bell and J. M. Bailey, the owner of the café where I worked, Coach Anderson was the third person in Eufaula who had a positive impact on just about every black youngster coming up. He was the only Native American coach I'd ever seen. Black coaches were about as common in our parts as caviar. The first time I saw a black coach on my side of the ball was at Oklahoma University when Lucious Selmon returned to assist Barry Switzer. We also had another black coach by the name of Wendel Mosely.

There were a lot of things that impressed me about Coach Anderson. He was a great motivator and teacher. He cared about his players as people, not just as a means to glory. He also taught us that appearance is important; taking pride in how you look telegraphs the kind of person you are to others. Coach Anderson was a sharp dresser, and as someone who had worn too many patched jeans, I was impressed. He was one of my role models. I even used some of my hard-earned cash to buy a leather coat cut like one he wore. And if that wasn't enough to cement a friendship, his wife made truly great coconut cream pies.

To this day, I think "Coach" is one of the most respected titles a person can hold. The mayor of a city in my district is a former football coach, and when we talk, I find myself wanting to call him Coach instead of Mr. Mayor because, for me, it bestows even more respect. That's certainly how I feel about all the coaches I played for over the years. Many of us would not be where we are today without the guidance and wisdom of a good coach along the way.

Sports were clearly the dominant force in my life during high school. I was a decent student, but I'll admit that sports were my main motivator. I had too much pride to risk standing on the sidelines on a Friday night, stripped of eligibility because I was failing a class.

Even though I could have paid more attention to my course work, I loved going to high school. Sports all year round, school dances, studying when I had to, and the friendship of a great group of kids that got along amazingly well. I liked to dress sharp during my high school years—as far as my limited budget would allow me to go. I was still working, but I'd learned my lesson about overspending for clothes. My idea of "sharp dressing" sends chills up my spine now. My favorite outfit at the time was a navy blue shirt that coordinated, 1970s style, with a pair of red-and-blue-striped bellbottoms. Since I know I wasn't on drugs, the only excuse I can think of is an overabundance of bicentennial spirit.

One Saturday night, just as I was about to go out on the town in this vivid getup, Grandma Mittie called and said she needed me to help her with something, so I hustled down to her place with my bellbottoms flapping. The "bells" were so big you could hear me ringing a block away. I had just a few minutes to spare before I was supposed to meet the boys. Mittie used to do ironing for some of the well-to-do folks around Eufaula. She could press a shirt or a pair of jeans well enough to put most professional cleaners to shame. She always took care of my shirts and pants, and they were as clean and pressed as anyone's in town. Well, when I arrived in her living room and asked what she needed, Mittie led me into her kitchen. There, smelling up the whole room, was a sinkful of fish she'd just caught out of the lake.

"I want you to clean and scale these for me, Junior," she ordered. Everybody called me Junior in those days—some still do.

"But, Grandma," I protested, "look what I'm wearing. I've got a date. This is my Saturday-night best."

"That don't make any difference to me. I need those fish cleaned."

For possibly the only time in my life, I refused a direct order from her.

"No, Grandma. I'm not going to ruin my clothes just to clean your fish."

"Well," she said, "don't bother bringing your ironing down here next week!" And she meant it.

Two weeks later, though, when I went down to help her with some other chores, she passed a stern eye over the wrinkles in my school shirt.

"If you want those things ironed," she commanded, "you'd better get them down here first thing in the morning." And just like that I was back in her good graces again. Truth be told, like that sinkful of fish, those pants of mine would have been better off at the bottom of the lake.

My rows with Mittie were few and far between, and as I grew older, the fistfights that highlighted my middle school years disappeared with bicycle riding and water balloons, to be replaced by football playing and girls. I was a pretty friendly cuss who could usually see something to like in almost everybody, and I soon found myself evolving into leadership roles within the school.

One of my most memorable and fun high school experiences had nothing at all to do with sports. In my junior year, I was one of several football players who had a fifth-hour study hall that started at 1:30. Sixth hour for us was the beginning of football practice. I'd like to be able to tell you I spent the study hall hour with my head buried in my books sopping up great literature or unraveling the mysteries of science. For most of us, study hall was an exercise in "gotcha." How far could we go before the teacher yanked our chains and shut us up?

All that changed the day Susie Moores strode into study hall. I had known Susie since I was 8 years old when her husband, Gary, was a Summer League baseball coach. They are a great pair and a perfect match. When they decide to do something, get out of the way! There's no stopping them.

Gary and Susie were two of Eufaula High School's biggest supporters. Still are. Along with Daddy and Uncle Wade, they used to cheer us on at all the games, even the away games. One year, we were in the state football playoffs, and I got knocked sideways and ended up with a concussion. It was Gary and Susie who took me to the hospital in Tulsa. Later, I didn't even remember the game, but I'll never forget their kindness and concern. They came to be like second parents to me, and a kid can always use an extra mom and dad. Given the amount of time I spent at their house during high school eating Susie's good cookin', I suspect they almost could have claimed me as a dependent.

When the study hall incident took place, Susie was the music teacher for Eufaula public schools, but she was itching to do something more. Being blessed with a lot of drive and a little blarney to go with it, Susie had buttonholed the high school administrators and talked them into giving her the high school chorus. When Susie took over, it was an all-girls glee club. She wanted to make it a mixed chorus. Now all she had to do was convince a school full of boys who thought singing was for sissies to "break the gender barrier."

Susie knew the boys of Eufaula High would not enroll in chorus without some friendly prodding. What we didn't know when she marched into study hall that day was that she had come for us—the football team.

A group of us were sitting together in the back, as usual. It made goofing off easier that way. Susie walked right up to us, but it was ol' J. C. who was in her crosshairs.

She looked me straight in the eye and said, "J. C., you need to be in chorus, not sitting here looking stupid at each other. If I can get you to join, then I can get the rest of the boys to follow." Susie understood that a leader knows where he's going and convinces others to go along, and she had tagged me for the job. "Beefing up" the chorus was the mission Susie gave us that day. Now, all we had to do was get the rest of the male population on board.

What she wanted us to do was lead. I understood that completely, and I liked the idea. I wasn't averse to having a place to exhibit my musical "talents," either. The little boy who liked to sing at the Sulphur Springs Baptist Church still lurked inside the bulked-up football player sitting in study hall that day. Listening to me bellowing in the bathtub or singing my way home from school, my sisters could attest that I had never kicked the habit.

At parties, I used to use an Afro comb as a mike and sing along with the Temptations. Bashful I wasn't. Some things never change. A couple of years ago, the United Way in Norman had a karaoke fundraiser, and the Watts "singers" did a great rendition of "My Girl," the Temptations' number one hit from 1965. By now, you probably think I have a repertoire of one song, but in my view, "My Girl" is the most beautiful song ever written. When we got to the line, "I don't need no money, fortune and fame," I pulled a few dollar bills from my pocket—Monopoly money—and tossed them in the air. Seeing a U.S. congressman throwing money around is nothing new, but doing it to a Motown sound got some attention.

Back in 1974, the football team taking up chorus got some attention, too. We should have warned Susie to be careful what you wish for. She soon found herself with 98 students in her fifth-hour chorus, and more than half were guys. During her sixth-hour chorus, when the football players were hitting the field instead of the low notes, she had another 99 kids, boys and girls.

We had a blast singing with Susie! We filled the band room, bumper to bumper. We loved our music, and so did the town: we were often asked to entertain. Before long, we were putting on concerts with drummers, guitarists, a whole band that played with us. Susie was smart enough to fill the repertoire with things kids would enjoy, like Beatles tunes and traditional folk songs like "This Land Is Your Land." It was great to learn that songwriter Woody Guthrie came from Okemah, an Oklahoma town no bigger than ours and only about an hour's drive away. And that he not only wrote some of this country's most loved songs, he also inspired Bob Dylan, Bruce Springsteen, and a whole lot of other musicians who have come on the scene in my lifetime.

We quickly became a troupe of traveling musicians singing in Masonic lodges and any other place where people would have us. It was great fun and a good way to keep kids out of trouble, too.

I don't know how many American kids have the kind of school situation today that we enjoyed back then, but at Eufaula High everybody knew everybody else. I sometimes think many of the problems we see in high schools today, especially the problems of violence, can be traced to the trend toward bigger and bigger schools. Large schools certainly have economic advantages, but I worry that we are sacrificing safety and socialization in the drive for "superschools" that provide everything but a sense of belonging.

Eufaula was tiny in comparison to most schools today, and with few amenities, but it had a lot of heart. Sometimes during football season, I'd walk between classes and hear someone holler, "Drop the bomb!" which meant a touchdown pass back then, not what we might think today. School spirit ran deep at Eufaula High.

Coach Bell once told me he had a kid come into class at the end of the year with tears in his eyes. Coach asked him, "What's wrong, son? You've got an A going for this class, and the whole summer's ahead of you."

"That's just it, Coach," the kid said. "I don't want school to end." Now there's a response every parent and teacher would love to hear.

I'm not suggesting that Eufaula didn't have a few troublemakers. I remember as a young boy running from a pair of brothers who beat up little black kids for sport when they could catch us. I suppose I should be grateful to the McGuire boys for developing my running skills as they tracked us in their truck, and I dashed between houses and fields to avoid them. In the end, however, I got the last laugh. They went to jail, and I went to high school and then OU on a football scholarship. Most kids in my hometown, however, were too busy being involved in school activities and part-time jobs and just having fun to get themselves into too much trouble.

Thanks to sports and other activities, Eufaula High was a school where racial problems were rare, but there were times when race was an issue off the playing field. By 1975, Dr. King's mighty words had raised the consciousness of the American people, and we were moving toward the "beloved community" that he talked so much about. But there were still unspoken rules in Eufaula.

In school, I had a lot of black *and* white friends, but after school, I generally hung out with my black buddies. There was no antagonism between the afterschool groups. We often went to the other side of the tracks to "fill up" at the Ric Rac or Jolly Bob's, and there was never any problem. Actually, we would have thought it far stranger to see a white kid on our side of the tracks hanging out in the pool hall or one of the markets—except for Pat Adams or Danny Kirby. We considered this pair a part of our pack—soul brothers, if you will. I still talk to Pat three times a week to this day and consider Pat and Danny dear friends.

Most of the white guys, however, thought there was nothing they would want in our neck of the woods. Everybody, black and white, spent a lot of time "draggin' main," cruising up and down the center of Eufaula—a short trip to say the least, maybe 3 miles. We'd greet our white friends as we rolled by, and they would do the same. We can thank sports, in large part, for the friendly relationship that we all shared.

Sports also caused me one awkward moment. Eufuala High had its own homecoming traditions. The team voted for its queen, and as captain of the football team, I served as king. My senior year, the queen was white, and for the first time, the school had to deal with an interracial royal couple. The rub came with one of the school's other traditions. After they were crowned, the king usually kissed the queen in the middle of the football field. There was some rumbling from those who hadn't been happy when

Coach Bell named me quarterback. It was as confusing for me as it was for everybody else. The big night arrived, but in the end, we didn't have the kiss that might have caused more problems than it was certainly worth. No big decision had been made. We just didn't do it. I think it was for the best, but it reminded me and others that as much as Eufaula had changed, we still had issues that none of us were certain how to handle.

My senior year, I was elected student body president, the second African American to hold that honor. Dewey Selmon had been the first. I was flattered and grateful to my friends, black and white, who had chosen me to lead them our final year. One of my responsibilities was to make introductions at school assemblies, and I discovered I enjoyed public speaking.

Every generation wants to leave its mark by making a few changes in the way things are done. I decided that Eufaula High was overdue for a real junior-senior prom. Up until my last year, the banquet and dance had always been held in the school cafeteria—not exactly the Ritz. We definitely needed a change.

So, once again, I collaborated with Susie Moores, Danny Kirby, Holly Chandler, Billy McNeal, and a few others to find a new home for the prom. After some discussion, and with options pretty limited in little Eufaula, Susie came up with the idea of moving the prom to the Fountainhead Lodge up on Lake Eufaula. We had a plan. A few of us met with the superintendent of schools, and he gave us the go-ahead—sort of. If we could raise the money, he told us, the board would consider our request.

All the kids got on board immediately. It wasn't exactly a tough sell, but the teachers were another story. They were not enamored with the idea. Some worried about kids drinking and driving on the narrow winding roads around the lake. Some were concerned that being supportive might jeopardize their jobs somehow. Susie and Perry Anderson, however, were solidly behind us, and both volunteered to sponsor the event. The students decided to police themselves and did a good job. Working together, we raised the money and decorated the Lodge to a fare-thee-well. We were ready to celebrate the end of the year and, for some of us, the end of our time at ol' Eufaula High.

But sometimes we are tested at moments when we least expect it.

On that day of so much excitement and anticipation, the stars crossed and produced tragic results. Steve Thompson, a junior, left school early that day to drive the 13 miles to his home. Excited to be attending his first prom, he said he wanted to get a jump-start on getting ready, and he still had to pick up his tux. Steve never made it to the dance. We got a call that there had been a wreck and our buddy had been taken to Tulsa.

No one knew what to do. Cancel the prom—or go ahead and hope for the best? We talked about it for some time, but finally we decided to hold the dance because so many people had worked so hard. The celebration turned into a wake when we got the news that Steve had died. I tried to make some remarks to the students. It was one of the most difficult things I've ever had to do, and I don't know if anything I said made anybody feel better. There is simply no way to explain the loss of a popular young man on what should have been one of the happiest nights of his life.

That year, the prom was a sad evening for all of us, but the class of '76 left a legacy. In the years to come, holding the junior-senior prom at the lodge became a tradition. That spring was a roller-coaster of emotions for me. I was showered with a host of honors and scholarship offers. Some of the great football schools in the country were recruiting me. Choosing the right one was a difficult decision, but my life those last few months of senior year was complicated by something even more serious. I'd finally run into Daddy's bear, and I had to face up to perhaps the most significant and difficult experience of my life.

I was strong and fast enough to outdistance almost every opponent on the football field, but I could not run away from my own selfishness. In the fall of my senior year, I fathered two children—both were girls.

Sports in most ways are good for body and soul, but when you're 18 years old, you don't always have the maturity to deal with the hero worship that comes with football fame. My name was in the papers, recruiters were singing my praises and offering me pie in the sky and my choice at JM's, too. My ego got the better of my judgment.

When I found out about the pregnancies, my first thought was, "This can't be happening to me. I know where I'm going." The life I'd been dreaming about since I saw Lucious Selmon race across the OU field suddenly seemed about to slip from my grasp. It was a selfish reaction that was quickly followed by terrible regret and guilt that I could have done this to two young women who deserved a whole lot better from me.

This is not an easy thing for me to acknowledge even 25 years later. It's difficult to elaborate on it without drawing people into the equation who probably don't want to be part of this book. They deserve their dignity and privacy because I was the one who should have known better.

Honesty compels me to discuss my early unwed fatherhood, yet decency also compels me to protect people's privacy. As much as I regret my foolish behavior and would warn any young man against making the same selfish choice I did, I am also grateful to the Lord for giving me those two wonderful daughters.

Today, they're all grown up—attractive, accomplished, and extremely special young women. This world is a better place for having them in it. I'm a better person for having faced up to my responsibilities as best I could, but fortunately for me, I didn't have to go it alone. During that extremely difficult period in my life, I discovered the true meaning of family.

Buddy didn't kill me as I feared he might. He was understandably angry with me, though. I was a child having children because I'd been heedless and irresponsible. But my family rallied around me and committed themselves to turning something that could have been tragic into a magnificent show of love.

Uncle Wade and his wife, Aunt Betty, adopted and raised one of my daughters as their own child, making something good come out of a bad decision. There were never two people with more charity and love in their hearts than Wade and Betty Watts, and I thank God for having put them in my life.

Frankie, the wonderful lady I married, enrolled in cosmetology school in Oklahoma City and our daughter, my other child, stayed home with our parents in Eufaula for about 10 months off and on until I was established in college and we were ready to get married. Frankie and I were lucky to have the support that helped us found a strong, lasting, loving marriage that in time was to produce four more children.

To this day, I am sorry for the pain I caused so many people, especially my wife, Frankie. So I'm grateful to her for bearing the burden of my selfish decision. But when telling the story of our lives, we must be honest about the bad along with the good.

We all make choices we're not proud of, but that spring, in 1976, I learned the hard way about personal responsibility. I didn't walk away from the turmoil I'd created, although I sometimes wanted to. And I'm glad that I didn't. I wouldn't have Frankie or the family I have today if I'd taken the easy way out.

But I learned something else in those last months of high school. I came to understand the meaning of unconditional love. Anyone in politics or athletics or entertainment knows that too many people who want to be your friend in good times disappear faster than a plate of Susie Moores's cookies if you find yourself in hot water.

I was up to my eyeballs in trouble, but my family didn't desert me. Frankie gave her love and her all to our daughter and to me. Uncle Wade and Aunt Betty were there to help me pick up the pieces of my life and give a loving and happy life to another. They didn't say, "You've made a bad choice. You're on your own."

Instead, they taught me the power of a strong family in overcoming adversity, and if I achieve anything in Washington that serves to strengthen the families of this country, I will go home to Eufaula a happy man.

It was in high school through good and bad times that my opinions and attitudes began to form. The scope of my vision broadened beyond athletics. I began to see myself as more than a football player. I didn't see the U.S. Congress in my future quite yet, or the role that faith would play in my life, but I knew I wanted more and began to believe I could make it happen.

It was time to choose the next path. I had my dream—Lucious Selmon had given me that. Now, I had to decide which college would give me the best chance to make the dream come true.

7

Never Read Your Own
Press Clippings

> A boy becomes an adult three years before his parents
> think he does and about two years after he thinks he
> does.
>
> Lewis Hershey

To some people in Oklahoma, Barry Switzer is close to a god. To others, mostly outside the state, he's a controversial man with a penchant for attracting trouble and troublesome players. The press likes to call him "the King"; I call him Coach.

I first laid eyes on Barry Switzer the summer after my freshman year in high school when the little town of Eufaula put on one mighty big celebration—Lucious Selmon Appreciation Day. We also honored Dewey and Lee Roy at the same time. The highway that leads from Eufaula to Oklahoma University was renamed Selmon Road, while the highlight of the day was a parade down Main Street with Lucious waving to the crowd from the back of a gleaming convertible.

What a day for Eufaula and for the Selmon family! Practically everyone in town was there to cheer on Lucious and his brothers. But the Selmons weren't the only attraction in the parade that day. Right behind

Lucious on a red flatbed trailer was the legendary Barry Switzer sitting on a pile of hay bales. What you have to understand is the degree of celebrity Coach Switzer generated in those days. No one in Oklahoma, including the governor and possibly even the president, was more important to more Oklahomans than Barry Switzer. Throughout the 1970s, Oklahoma was the top college football team in the United States as Coach Switzer led his Sooners to national championships in 1974 and 1975, his first two years as head coach. Ten years later, in 1985, he would repeat his national championship performance.

With the nearest professional teams the Dallas Cowboys and the Kansas City Chiefs, for Oklahomans, then and now, following high school and college football borders on religion. Watching the pros duke it out was fine for Sunday, but on Friday nights and Saturday afternoons, local and college athletics were the heartbeat of community life in little towns like Eufaula all across Oklahoma. And back then Barry Switzer was king of the hill—the constant presence on the OU sidelines, the guy whose quicksilver football mind and coaching instincts had kept the Sooners ranked among the best in the nation for years.

By the time Coach Switzer came riding down Main Street in Eufaula, looking surprisingly dignified as he was pulled along by a big Ford tractor, he was already becoming the stuff of legends in Oklahoma, and for one 15-year-old high school football player standing in front of the pool hall that day with stars in his eyes, seeing Barry Switzer was a heart-stopping experience.

My buddies and I had gotten to Main Street early that afternoon to make sure we had one of the best spots. The pool hall was a perfect place to get a "50-yard-line view," and as Lucious cruised slowly by, I caught his eye.

"Hey, Lucious!" I hollered and waved.

Instead of answering me, Lucious turned around and yelled back to Switzer, "Hey, Coach! See that guy over there?" He pointed at me standing in the doorway of the pool hall.

"You mean that one?" Switzer asked, looking straight at me.

"Yeah. He's going to be your quarterback someday. His name is J. C. Watts."

The coach grinned and yelled, "What's J. C. stand for?"

"Julius Caesar," Lucious hollered back.

"Hey, J. C.! Are you gonna come and play for me?" Coach asked.

I must have grinned from ear to ear. Maybe just maybe I really could become Barry Switzer's quarterback! For Coach Switzer, I'm sure it was just a harmless exchange on a day when everyone was celebrating. What were the odds that some country kid standing in a poolroom door with the breeze whistling through his 'fro would get past all the obstacles and find himself signed on with one of the great American football dynasties—let alone starting at the highest-profile position?

It might have been a long shot in a lot of people's minds, but for me, it was a message from Mount Olympus. The dream I had been carrying around since Thanksgiving Day 1971 suddenly seemed within my grasp. And once again, it was Lucious Selmon who made it happen.

I would have packed my bags that day, but I was getting ahead of myself. I still had another couple of years of high school ball to play before I would remind Coach Switzer of our "agreement" that day in Eufaula. But my decision to attend Oklahoma wasn't as easy as I thought it would be. By my senior year, recruiters from dozens of colleges and universities had made the trek to Eufaula to check out my abilities and me. What followed were offers from some of the best institutions of higher learning in the country—schools where I could get both a good education and a chance to play world-class football.

I've always thought football recruiting was a little like negotiating a royal marriage. Dynasties rise and fall on the strength of the matches made in locker rooms and restaurants and living rooms across America. Coaches, college officials, athletes, and their parents involve themselves in a kind of strange mating ritual while evaluating each other's strengths and weaknesses, weighing educational opportunities and financial scholarships, and calculating the odds that the alliance will produce an outcome that will satisfy everyone involved. Sometimes, it's a love match, as OU was for me; other times, it's more a marriage of strategic advantage.

Throughout that recruiting year, however, it often felt like I had almost too many choices. The pressure on me to pick the right school was enormous. Despite the promise I'd made standing in the pool hall doorway a few years before, I understood the wisdom of considering other schools along with OU before settling on one. The recruitment was hot and heavy.

Some schools recruited my parents even harder than me. A few recruiters even offered improper inducements, the kind that can be difficult for a poor family to turn down. But Daddy and Mama knew the difference between right and wrong, and there was never any question but that we would follow the rules. At least I ate pretty high on the hog that year as I was treated to some of the best dinners in Eufaula and on some of the campuses I visited.

One of the strangest recruiting episodes happened one afternoon at Paul's Chicken Corral. It was my senior year, and a Notre Dame recruiter was in the state to scout out the local talent. Oklahoma has a reputation for producing outstanding football players, and this recruiter was traveling the state looking us over. Apparently, as he drove from town to town, he kept hearing about "this Watts kid," and one afternoon when he discovered he was close to Eufaula, he decided to take a chance and drop by.

I just happened to be at the Chicken Corral getting some gas when this stranger pulls in and rolls down his window. He introduced himself to me, and said, "I've been hearing a lot about this Watts kid in Eufaula. They say he's some football player. Can you tell me where I can find him?"

I laughed, and said, "Yes, sir. I sure can. I *am* the Watts kid." His mouth dropped, and I'm sure we both thought, "What are the odds?" We had a friendly conversation, but Notre Dame didn't pan out or offer a scholarship.

Then, one afternoon in our living room, after an endless parade of suitors had courted us over the months, good old Barry Switzer—as confident and cocky as he could be—told my parents, "Other schools are going to offer J. C. money and lots of other things. We're not going to give him a damn dime!"

I understood what Coach meant, but I could see that his frank language got Daddy's back up a little bit. I don't think Buddy ever quite got over Coach Switzer's attitude that day. In fact, for years, when the subject would come up, he would say, "I just kick myself that I didn't tell him 'The Watts family is a poor family, but we're not a begging family. We're not asking anybody for anything.'"

The University of Oklahoma's greatest selling point, besides Lucious Selmon, was also its biggest drawback. It was a dynasty, the winningest college football team in the country and, as many people warned me, a place where a kid like me, with all my dreams of glory, might get lost in the intense competition of football at that level.

My daddy was leaning toward Oklahoma State for exactly that reason. He was concerned that OU was going to be a tough road to walk. He made a good point, but I just couldn't get that picture of Lucious Selmon on television out of my head, so I turned to my good friends, Gary and Susie Moores, for a second opinion. They were all for Oklahoma, for its academics as much as its athletics. All the Selmon brothers told me they held the deepest respect for Barry Switzer, and their opinions meant a lot to me, too. In the end, I knew in my heart I had never run from a good fight, and I wasn't about to start now. I also knew this difficult decision was mine alone to make, and I chose OU—where my heart was all along. I had a chance to make my Thanksgiving Day dream come true, and I was going to take it.

The Mooreses had just built a house that was so new the lawn wasn't much more than a mess of muddy wheelbarrow ruts, but they had graciously offered their home for the signing ceremony, which took place on February 7, 1976. Sitting in the Mooreses' living room, I chose my own path—one that would take me just a little more than a hundred miles west to Norman, but it might as well have been halfway around the world. I was leaving Eufaula, which had been my safe haven for 18 years, for the bright lights of Big Eight football and the academic challenges of a first-rate college. I was scared to death.

When I arrived at the University of Oklahoma, my first months were pretty rough. The academics were tough, the football was tougher, and I missed Daddy's hollering and Mama's cooking. I even missed the hard work.

It was a difficult period for me. Looking back now, 25 years later, I realize I was beginning to change from a boy to a man—never an easy transition for anyone. But that's not what I thought then. Only weeks after leaving home, I came to the conclusion that I'd made a big mistake.

Back in little Eufaula, I had been the classic big fish in a small pond—All-State and All-American football star, team captain, and student body president. I was voted Eufaula's best all-around male student, and Back of the Year for the whole state of Oklahoma. I just assumed I would arrive at OU like a reigning gladiator entering Rome. I'd be starting quarterback my freshman year, of course, and my playing prowess would eventually earn me a spot in the NFL and the millions of dollars that go with it.

After this long stretch of hearing how good I was, how I was sure to become All Big Eight and win national championships to boot, it's no wonder I walked into the locker room that autumn afternoon in 1976 and suited up for my first college practice acting a little like the only rooster in the henhouse.

Abraham Lincoln used to tell the story about a man who asked to be appointed to an important post overseas. When he was turned down, the man asked for a more modest job, but again Lincoln said no. Finally, he asked to be named customs inspector.

As Lincoln explained, "When he saw he could not get that, he asked me for an old pair of trousers. It is sometimes well to be humble."

Well, I got my share of humble pie and then some my freshman year.

Just before heading for the field, my eyes focused on a piece of paper tacked on the bulletin board. I realized it was the team's depth chart, and I charged over to find my standing. Then reality hit. I felt like I'd just been blindsided by Lawrence Taylor. There were eight quarterbacks on the OU roster that year. Eight! I was devastated, and nearly cried when I saw my name at number 7 right next to the bottom.

Then I remembered everybody's warnings about OU.

"They'll just see you as a number," someone had said. Yeah, I thought, number 7. A voice in my head began echoing the doom and gloom I'd heard from the naysayers: "Maybe they were right." And suddenly that defeatist idea started to have powerful legs.

On top of that cold splash of humility, I was more homesick than I could have believed possible. I was a small-town kid overwhelmed by the size of OU with its more than 23,000 students. The whole town of

Eufaula was only a couple of thousand people. This was culture shock in the extreme for a country kid missing his mama's cooking and the family and friends who had always provided a firm support system.

I was also finding the academics much more difficult than I expected. I wasn't one of those student-athletes who didn't care whether I got the degree or not. Pro ball wasn't my only goal, but my determination to succeed both in athletics and academics meant working harder than I ever had in my life to keep my grades up to par.

I suppose I wasn't going through anything much different from the challenges that derail so many local-hero athletes when they arrive on a big campus like OU, but that didn't make it any easier for me at the time. The cocky J. C. Watts who yelled at Barry Switzer from the poolroom doorway had all but disappeared, replaced by an insecure, lonely, scared kid wondering if he had made the wrong choice.

It wasn't long before I was burning up the phone lines home and neither Mama nor Daddy approved of long distance unless it was a dire emergency. Well, from my point of view, it was. I told Mama, "I'm lonesome and I just plain don't like it here."

Mama and Daddy would say, "J. C., don't give up. The first year is always tough, but it'll be worth it in the end." Good advice from people with my best interests at heart, but I didn't have the heart to stick it out.

Gary and Susie Moores drove to Norman for every home football game that year. We'd get together after the game, and Susie would always have a care package ready for me filled with her great homemade goodies. Some people need a doctor's prescription to get rid of depression. All I seem to need is a good dessert or two. But once the Mooreses had driven home and the goodies were history, I soon started feeling sorry for myself all over again.

I managed to hang on, though, and thanks to a little luck, a couple of the contenders for quarterback moved to other positions by the start of the season. I had moved up to third team, and I made the traveling squad. That should have been incentive enough to keep me in line, but when I played a total of six downs my freshman year and another six in the postseason Fiesta Bowl, no one could have felt sorrier for himself than I did at that moment.

If I wasn't going to get to play, I reasoned, I might as well pack up and go home. And that's exactly what I did. I went home to Eufaula to wrestle with my own feelings. Luckily, nobody knew about my exit but me. By the time Sunday rolled around, and it was time to drive back to campus, common sense had returned, at least temporarily. Just being at home for a couple of days recharged my batteries, and I decided to give OU another chance. Wattses weren't quitters.

Unfortunately, that wasn't the only time my immaturity got the better of me. The second time I decided to chuck OU, football, academics, and my future—not necessarily in that order—was a few weeks before spring practice was to begin and just a few months before Frankie and I were to be married.

I can remember my middle-of-the-night "escape" as if it were yesterday. It still burns in my mind. I had been wrestling with my anger and frustration for days, having learned that another year of inactivity was likely. Finally, late one Thursday night, the dam broke. I collared two of my buddies in the dorm—Freddie Nixon, a wide receiver out of Florida, and Darrol Ray, a defensive back from Killeen, Texas.

Neither one was much in a mood to do anything but sleep, but I needed some moral support, and I hustled them both up.

"Guys," I told them, "I'm quitting, and I could use a little help getting my stuff down to the car."

I don't think they took me seriously at first, but it didn't take them long to realize that I was leaving with or without their assistance. Of course, I could have carried my own bags down to the car, but I had a need for speed. Once I'd made the decision to quit, I couldn't get off the OU campus fast enough. I'd never faced real personal defeat before, and I knew if I didn't get out of that dorm quickly, I might not make it at all. I didn't know what I was going to do; I just knew I couldn't stay there another moment.

Darrol, who has remained a good friend, kept telling me how nice and warm his bed had been before I'd gotten him out of it.

"You know, Watts," he said, "after all this, damn it, you better not come back!"

We packed up the car faster than a quick count, and I was on the road home to Eufaula. My first feeling was one of great relief. The decision was made—finally. Whether it was the right decision was an issue I wasn't ready to focus on just yet. I began thinking about a plan. I'd just transfer to Texas Tech or Oklahoma State or UCLA, my other options when I had been a senior on the recruiting circuit. Unlike Coach Switzer, *they* would appreciate my talents and would be happy to have me. I was convinced I'd be the starting quarterback at whichever school I chose, not realizing that the grass really wasn't greener anywhere else. Or that Coach Switzer himself had some big plans for me.

What I probably needed was a swift kick in the pants to get me to buckle down, get my head right, and assume personal responsibility for myself and my future. The Lord gives us what we need; we have to work for what we want. But playing the victim was a whole lot easier.

Former Indiana basketball coach Bobby Knight was a controversial figure but he knew a lot about winning. He once said, "The will to win is not nearly as important as the will to prepare to win." That's good advice for football, too. In fact, it's good advice for anyone pursuing a dream

I was being tested. Did I have the right stuff to be the quarterback of one of football's great dynasties? The answer at that moment was no. I wasn't willing to pay my dues, and that shallowness and lack of focus nearly sabotaged my own dreams.

About halfway home, I decided I needed some time alone before facing the "dragon" back in Eufaula, so I checked into a motel in Muskogee and spent the night. On Saturday, I headed home to tell Daddy and Mama.

"You know, Junior," Daddy said when I gave him the news, "I'm not going to try to influence you one way or the other. You're at an age now where you have to make these decisions on your own."

I thought I was home free, but then he went on.

"But I can tell you this much. If what you're doing was easy, everybody would be doing it."

Daddy's words were like a bolt from heaven. Everyone wasn't doing it. I was trying to achieve something extraordinary. Why did I think this was going to be easy?

"I'm trying to be the starting quarterback at the University of Oklahoma, which has been one of the top schools for football in the country year after year," I thought. "That's something only one person in the whole world can do."

Not only did Oklahoma have two recent national championships to boast of, but it had won outright or tied for the Big Eight Conference

Championship every year I was there. During the 1970s, OU produced a Heisman Trophy winner, two Outland Trophy winners, and an astounding number of pro players.

This was the level of talent I was competing against. What I had to decide was clear: Did I have the "right stuff" to go up against some of the best players in the nation and accept the outcome, win or lose? In other words, I had to grow up, give it my best shot—no excuses—and live with the consequences.

Once again, Daddy's wisdom, derived from a lifetime of personal responsibility, gave me the guidance I needed. My daddy never played football, but his own life experiences had given him plenty of understanding about the challenges and difficulties I was facing. As a young man, he had been a boxer. He knew what it meant to try to be the last man standing. He knew it was easier to just leave the ring, but he also knew that nothing worth winning is ever achieved without a few knocks.

And after years of scrambling for jobs, no one understood competition better than Buddy Watts. Most important, he knew that there are times in every man's life when reaching for your dream forces you to find bedrock, to look deeply into yourself and summon the courage to do your best. That's all Buddy Watts ever expected of me growing up, and that's all he expected of me that long weekend in the spring of 1977.

While I was talking to Daddy, Gary and Susie Moores were talking to Lucious Selmon. Word travels fast in a small town, and my unexpected arrival home was news. Lucious was an assistant coach for Oklahoma at the time, the first black coach I'd had and a friend as well.

When the Mooreses told him I'd quit and was home in Eufaula, Lucious's next call was to Barry Switzer. Coach Switzer called me at home and said, "Come back, and let's talk. After that if you decide you want to go, I'll give you a release."

That's a man with a lot of confidence in his own ability to lead a team and lend a hand to a confused player when he needs one. He had invested a lot of trust in me. Coach Switzer could have refused to give me a release and effectively ended my career, but he gave me the choice.

With a big lump in my throat and still weighing Daddy's wise words, I got back in my car and headed west to Norman and my meeting with Coach. In his office the next day, Switzer looked me right in the eyes as he said, "J. C., if you'll stay, you'll play. You've got a future here."

But Coach was straight with me.

"You didn't get to play a lot last year, and this next season I'm going to red-shirt you. You won't get to play at all. Then you'll come off red-shirt as backup. After that, you'll start for two years."

That was the deal—plain and simple. He could have tried to sugar-coat it, tell me what he thought I wanted to hear, but he didn't. He was honest with me. Not every player may like Barry Switzer, but no one will ever accuse him of not being straight. That was the real beginning of my lifelong friendship with Barry Switzer. Like Coach Bell, Barry Switzer's influence on my life was enormous. Over the next four years, his word proved to be as good as gold, and he kept every promise he made to me that day.

Ralph Waldo Emerson once wrote, "Our chief want in life is some-body who will make us do what we can." Coach Switzer was that kind of person for me. Coaches have it tough. All freshmen think they ought to be playing right away. I certainly did. But what separates winners from losers in any high-stakes environment is the mental game.

A heck of a lot of people have the physical skills for collegiate and even professional athletics. Finding people with the mental fortitude and leadership skills is another matter. You don't have to be born with them, but you must be willing to work for them.

I can't say I had a terrible attitude, but it certainly wasn't the right attitude. I thought I knew better than the coaches. I had become selfish, putting my wants ahead of the team. That selfishness led to discouragement, and it was a quick trip to envy. I believed I was the victim of a system that really didn't give a hoot about me.

As a student-athlete, I could not have been more wrong. Most students are on their own when they go off to school, but the OU athletic department was like a second family for us right there on campus to help with studies, keep us focused, and be there when we had problems. At first I didn't appreciate that relationship. When I became a victim, my vision was so clouded I couldn't see the opportunities right in front of me or the people willing to do anything to help me take advantage of them. Whether I was right or wrong about the motivations of the coaches and tutors and administrators didn't matter that spring of my freshman year.

By embracing the role of victim, I was sandbagging the one person who could do the most to turn things around—myself.

I learned some hard lessons that year, none more important than the truth in the old adage that some things in life are worth waiting for, and the ability to develop patience is one of life's most important tools for ultimately getting what you want. It's a little like a diamond cutter working an uncut stone. It may take a hundred careful taps to fracture the stone in a way that leaves the most brilliant and valuable diamond possible, but the first tap is just as important as the last in achieving the final goal.

That became my reality. I had to work harder and without recognition. I had to trust both people and a process. I had to learn that all suffering is not defeat but a shaping experience, and I came to understand the important link between actions and rewards. Unfortunately, in today's cult of victimology, that link is missing in the lives of too many children who don't have Buddy Watts or Barry Switzer to straighten them out when they veer off course.

By the time I left Coach Switzer's office, all my problems weren't solved, but I was ready to change my life, and I did.

That summer, Frankie Jones, the little girl I'd met at Booker T. Washington Elementary, became my wife, and we settled down in a little apartment near campus in Norman. Frankie had been going to school in Oklahoma City, where she was also raising our daughter. We had been seeing each other throughout my freshman year, trying to decide what was best for us and for our child. We wanted to do the right thing, but we just weren't sure what that was.

Finally, one evening after we'd gone to a movie, we were sitting in my car, and I asked her to marry me.

I really think I caught her off guard. All she said was, "Are you serious?" But eventually she said yes, and we were married in Eufaula in the teen hangout Daddy owned. Uncle Wade did the honors.

Daddy and Uncle Wade were known far and wide as two of the funniest men around. Those two together could keep a funeral in stitches, so a wedding was easy pickings. Daddy led the charge.

Frankie and I stood nervously before Uncle Wade with the family and a few friends all around us as we made a serious commitment to each other and to God. I had already recited my vows when Uncle Wade turned to my bride at that most solemn of moments in any wedding and asked, "Frankie, will you take this man for richer or poorer?" But before he could finish, Daddy piped up for all to hear: "Mainly for poorer."

It brought the house down. He and Uncle Wade just howled.

At least, Frankie and I were off to a happy start. We were both very young and had some adjusting to do like most couples. Frankie had been raised in a home with a loving mother but no father, so she had to learn the role of a father in a family. I had to learn how to juggle the responsibilities of fatherhood with studies and football. And we both learned that marriage is a journey, not a destination. After 25 years, our journey continues.

Both of us grew up that first year in many ways, but one of the most significant was our discovery of the importance of faith in our new life together. With all the pastors in my family, churchgoing was a ritual for us. But faith isn't about attendance—it's about building a personal, intimate relationship with Christ that directs your life decisions. Together, Frankie and I found our faith, and in doing so, we chose another path that set us apart from many of the single college students around us. I began to do some lay preaching and speaking for the Fellowship of Christian Athletes. Given my gene pool, I suppose it would have been swimming upstream *not* to at least have thought about preaching. To earn some extra money, though, I also sold vacuum cleaners and Amway products, and in the summer I worked construction jobs.

That year, I also became much more focused on my development as both a student and an athlete, not because someone was watching over me (although Coach Switzer did have a gatekeeper by the name of Port Robertson assigned to keep the players in line), but because I felt a personal responsibility to become the best I could be. For many student-athletes, the emphasis is almost totally on the sports side of things. For others, it is a combination. For Dewey Selmon, his major, philosophy, was just plain strange. He was once asked about his odd course of study.

"Philosophy is just a hobby," Dewey joked. "You can't open a philosophy factory." But both Dewey and Lee Roy Selmon took the words "student-athlete" seriously.

Unlike Dewey, I majored in journalism with the idea of going to law school and doing sports broadcasting at some point in my life. I never lost my dream to be a professional football player, but I listened when Coach Switzer told me one day, "J. C., your academics will take you much further than your athletics." And he also talked to us about the importance of being good citizens as well as good athletes.

Every college player harbors the notion of pro ball at one time or another, but I got smart enough to know it was foolish to put all my eggs in that basket. I knew of too many guys who didn't study and then blew out their knees or had some other career-ending injury. They left OU and other universities with nothing more than memories. I had no intention of letting that happen to me. And thanks to Barry Switzer, the OU coaching staff, and Port Robertson, it didn't.

I did a lot of growing up over the next couple of years and found myself the last week of April 1979 in the middle of spring training and studying for exams. That weekend, I was scheduled to play in the annual varsity-alumni game, which traditionally closes out training camp for the season. Frankie, however, was involved in something a little more important: she was giving birth to our second child, Jerrell. I can still remember sitting with her in the hospital, my books piled beside me as I studied for my tests, all the while holding a cup of ice chips for Frankie.

Like most men who watch their wives in the throes of labor, I felt just horrible, although like most wives, Frankie would be quick to point out not as horrible as she did. This was the first time I had seen labor in all its glory, and secretly, I was congratulating myself on being born a man—as if I had anything to do with it. I think there's some truth in the old saying that if men had to give birth, our overpopulation worries would be over. As hard labor approached 36 hours, I did get the cue from my wife that it was time to put the book down and pay attention. Not long after that, Jerrell arrived in the world, and Frankie and I experienced a big—but joyful—change in our lives.

Becoming the father of my first son wasn't the only change in my life that year. A little more than four months later, on September 15, 1979, I suited up as the starting quarterback for the Oklahoma Sooners. The dream that began on Thanksgiving Day 8 years before had finally come true.

I was following one of the best quarterbacks ever to lead Oklahoma, Thomas Lott, who had started in the position for the previous 3 years. He had been a wizard at the complex Wishbone Offense that Oklahoma University ran, and he left very big shoes to fill.

The Wishbone is all about options, and the skills of the quarterback, especially mental skills. You start the play, read the defense, and make spur-of-the-moment decisions about the most effective way to move based on that read. It requires enormous preparation on the part of the quarterback, who must know the other team as well as his own, if not better.

Done right, the Wishbone is sheer poetry. Done the least bit wrong, it can be an epic disaster of mistiming, missed assignments, and missed opportunities. When my time finally came, I simply couldn't execute the offense as well as Lott. I can't think of anything else that better prepared me for a life in politics than my first two or three games as a starter.

Although we won those games, we didn't score 50 points or blow people out of the water the way an Oklahoma team usually did. The fans had higher expectations for the championship Sooners, and we didn't meet them, and for the first time in my life, I got booed on the field. It was a sobering experience.

My first reaction to being booed and criticized wasn't very diplomatic. I told a reporter, "You've got some jerk sitting in the stands that lost a six-pack of beer because we didn't cover the point spread, and he's upset?" That was a shortfall of mental toughness talking, and that's one thing Barry Switzer and Galen Hall, my quarterback coach, didn't tolerate.

Coach Switzer called me in and said, "First off, remember you know more about football than 99.9 percent of the people you're going to meet. You're the authority, so don't get sidetracked by other people's opinions."

He went on to say, "You stay focused. Don't pay any attention to the boos. Don't pay any attention to the fan mail. I let my secretary read my mail. When it's good, she gives it to me. When it's bad, she throws it in the trash." Over the years I've found Coach Switzer's words to be true: Listening to those who carp on the sidelines is a waste of time and focus.

Years later, shortly after I arrived in Washington, Newt Gingrich tried to give me some friendly advice about developing a thick skin in politics. Taking criticism was second nature to Newt, but I was the new kid in town, and he was giving me a heads-up.

I appreciated his concern, but I told him, "People being critical probably doesn't affect me as much as it does most members of Congress. I've been booed plenty before and often by my own fans." And that was an understatement.

I should have passed Switzer's advice on to Newt: Never read your own press clippings. I don't.

In the third game of the season, I seemed to hit my stride, and we beat the University of Tulsa on our home field. Unfortunately, we still had Texas to go, in Dallas. That game was the worst of my life—high school, college, or the pros.

You have to understand that back then, when Oklahoma played Texas, in reality, it meant very little. The game in 1979 had no bearing on their conference championship or ours, and it didn't affect Cotton or Orange Bowl bids. The truth is, we just don't like losing to Texas. When the Sooners take on the Longhorns, we're playing for state pride and a recruiting advantage, and that's enough to make it one of the biggest college football games of any year.

Fresh from our victory the week before, I was convinced we had turned a corner and were now on a direct path to the kind of performance Sooner fans had come to expect. When I'm wrong, I don't do it halfway. That day, I managed to single-handedly fumble the ball four times and throw three interceptions—seven turnovers all by myself. It's amazing that we only lost 16–7. The memory of that game still makes me cringe.

Coach Switzer said, "It has been my experience that the fastest man on the football field is the quarterback who has just had his pass intercepted." He could have been talking about me that disastrous afternoon. At least he didn't take Woody Hayes's approach to screwups. Gary Dulin, who played pro ball with me in Canada, used to keep me laughing with his Woody Hayes stories.

In Dulin's freshman year at Ohio State, his team lost to Missouri in a big upset, and he'd sat on the bench for the whole game. After the embarrassing defeat, Gary dragged himself back to the locker room and started taking off his equipment, throwing it angrily into his locker as he muttered, "If they're not gonna play me, they ought to let me go home. I'm gonna go to another school."

The coach stood there totally composed, never saying a word but taking it all in. Then, just as Gary started to take his shoulder pads off, when he was really vulnerable, Woody jumped him—all 285 pounds of him. Our postgame locker room was no fun, but at least the coach didn't physically attack me. Of course, nothing he could have said or done could make me feel worse than I already did.

After our disastrous performance, I came out of the locker room, and I remembered that a friend of mine was in the hospital in Dallas, where we were playing. So I fished my pockets looking for some change to give him a call. I thought I could cheer him up a little bit, and maybe he could do the same for me. After checking every pocket, I came up empty.

Then I saw one of our fans standing off to the side. He had on a red hat, a red coat, red tie, red everything. The guy even had OU imprinted on his front teeth. I was the quarterback, and I wasn't *that* team-crazy. So, I thought, surely this guy will give me a quarter to call my friend.

I walked over to him and introduced myself, "Sir, I'm J. C. Watts, and I've got a friend in the hospital here," I told him. "I just realized that I don't have any change. I was wondering if you could give me a quarter to call my friend?"

He reached into his pocket and pulled out some coins, and then lowered the boom: "Here's two quarters," he said. "Call *all* your friends."

We rallied after the Texas game, and with seven straight wins and a victory over Nebraska for the Big Eight title, we ended up 11–1 for the season. That record earned us a trip to the Orange Bowl where we beat Florida State in a great game. I was awarded Most Valuable Player, which took the sting out of the boos earlier in the season.

After the game, I was sitting in the lobby of the hotel recapping the season for a sports reporter. He was kind in praising me for coming back from so much early-season adversity and for proving my critics wrong.

Finally, he told me, "J. C., I've gotten some rough letters from folks. They don't say you're a bad quarterback or that you can't do the job. Some just say they want a white quarterback. Period."

"Not a whole lot I can do about that," I said. The coach wasn't concerned about my skin color, and that was all that mattered. He didn't care if I was green. If I could read the defenses and move the ball, he was going to play me. Everything else was nonsense and irrelevant.

In 1948, it had taken the legal talents of Thurgood Marshall and the Supreme Court to open the doors of OU's law school to Ada Lois Sipuel, a black woman. She sat alone in class behind a "coloreds" sign and was forced to eat in a special section in the cafeteria so she wouldn't mix with white students. My uncle Wade was part of the group that fought for Sipuel's admission.

Nearly 30 years later, it was the University of Oklahoma courting his nephew, not the other way around, but that didn't mean all hearts had changed. The race issue was an undercurrent at times during my tenure at OU, but it never became a problem. When that reporter brought it up, the grounding that Coach Bell had given me came into play once again. While he had taught me the game of football, he also taught me how to deal with people who could not accept those who were different from themselves. Coach Switzer's attitude mirrored Paul Bell's.

Going into my senior year, we were flying high, convinced we had a great chance to win it all that year. There was no rookie quarterback to break in. All our principal offensive players and most of our defensive players were back from the previous year. We had practically made space for the championship ball in the trophy case. Then reality paid a call.

After four games, our record was 2 and 2. John Elway and the Stanford Cardinals beat us on our home field, and we lost to Texas as well. I couldn't seem to do anything right and neither could anyone else. My instincts failed me, and my physical skills in which I had always had complete confidence just weren't clicking. But there was no sympathy for ol' J. C. that year.

Back in the 1970s, that kind of win-loss record after four games was nothing short of sacrilege. Once again, a quarterback controversy erupted. The booers filled the stadium and rained abuse down on me. Some of the sportswriters weren't much better, and I was just as hard on myself.

The coach always knew how to handle his players. He called me in and said, "There's only one way to stop the boos—with your performance. We need to refocus."

And we did. A couple of weekends later, future NFL Hall-of-Famer Lawrence Taylor and the North Carolina Tarheels came to town with the number one defense in the country. We blew 'em out 41–7. Taylor was one of the best linebackers I'd ever play against, but we were ready for him. Our offensive strategy was to have someone block him on every play even if the action was going away from him, and that's exactly what happened. He became a nonfactor, but more importantly, we found our groove again.

We ended up going 10 and 2 that season, including a come-from-behind win over a powerhouse Nebraska team. Beating Nebraska was always sweet. Unlike Texas, we knew when the season started that the Nebraska game would be the big one to determine the Big Eight Championship and the Orange Bowl bid.

The rivalry between Oklahoma and Nebraska has always been in a league by itself. Jim Walden, a former Iowa State coach, once described Nebraska this way. "They're big. They're strong. They're fast. Their mothers love them . . . and they'll kill you."

I couldn't say it better myself. Nebraska is always tough, and the team to beat. It was a thrilling game and one of OU's most memorable victories. But our biggest game was yet to come—Florida State in the Orange Bowl. They were 11–0 that year, and came into the game ranked number two. We were ranked number four.

Playing in the Orange Bowl is a little like swimming in a three-story fishbowl in the middle of an earthquake. A hundred thousand screaming fans, TV cameras, loud music, spectacular half-time shows, fireworks, and more pressure than a ripe volcano. Focus is the name of the game, and this Orange Bowl was to be no different. It was a hard-fought battle from start to finish. With 2 minutes and 32 seconds left in the game, we were trailing 17–10. At a moment like this, a quarterback must do more than call plays. More than ever, he must lead, tell the team where he wants to go, and convince them to go there with him. Everything was on the line when we began what we knew was our final drive. We literally marched 65 yards down the field to the 17-yard line with about a minute to go. It was third down, and we had to decide whether to run or pass. The stands were thundering with cheers from our side and screaming from the other. Our adrenaline was pumping overtime. I'm sure Florida State's was, too. As Steve Rhodes, our split end, went wide right, I rolled out and threw the ball to where only Steve could catch it or miss it—low and outside. Rhodes reached for the ball. Touchdown! The stadium went wild but the game wasn't over.

With the score now 17–16, simply kicking the extra point would have brought us home with a tie and plenty of honor. Without hesitation, Coach Switzer made the call of a career: Go for it!

Because we were known as a running team, Florida State expected us to go to our strength. But we opted for a surprise play, and I passed to Forrest Valora, our tight end, for the two points. Actually, we only had one 2-point conversion play called 741 to the right side or 759 to the left side. We'd fake an option run to the tight-end side. The tight end blocks as though it's a run. The quarterback fakes an option run, steps back and, hopefully, the tight end is open. Forrest was wide open.

We had cinched another Orange Bowl victory as a team; and once again, I was named MVP. I was proud to eventually be inducted in the Orange Bowl Hall of Honor for my 1980 and 1981 performances. It was a glorious night for all of us. More important, Oklahoma finished its season as the number three team in the country. The dynasty and tradition continued.

Through all the boos and the strains on my emotions earlier in the year, I followed Coach Switzer's lead and he never led me astray. In football, when things go well, the head coach and quarterback get more than their share of the glory. But when things go badly, they get more than their share of the criticism. Coach Switzer could easily have bailed out on me when the going was tough early in the season. Everyone would have understood. He could have simply told me, "You're backup now."

But that wasn't his style. Instead, he came out publicly and told the press that J. C. Watts was his quarterback, win, lose, or draw. His willingness to take heat when my performance wasn't up to speed meant a lot to me, but Coach Switzer was always loyal to his players.

Bud Grant, the great Minnesota Vikings former head coach, said, "A good coach needs three things: a patient wife, a loyal dog, and a great quarterback—not necessarily in that order." Well, Coach Switzer had a terrific wife, a mean Siamese cat, and me. One out of three ain't bad. I know about the cat because at one point after Frankie and I were married I became the Switzers' baby-sitter.

He used to joke in speeches to fans that he paid his baby-sitters a hundred bucks an hour. I can attest to the fact that I didn't make anything close to that, but Coach always had a great sense of humor, and he never took himself too seriously. I remember one game with Colorado. The team was unusually wound up the day before, and the coach knew it. He broke the tension by strolling onto the field in a fur coat with a big cigar in his mouth.

I think most of his players would tell you that Barry Switzer was a transparent kind of coach—what you saw was what you got. He told you straight and was a great motivator because of it. I remember seeing a highlight film when he was named as head coach. There he stood in a plaid suit with longish hair, and I'll never forget what he said: "I'm a fighter. I'm a competitor, and I'm a winner. We're family."

And like a good father, Barry Switzer knew his kids. He was a great judge of talent and understood the personalities that went with it. Barry and his staff seemed to have a great ability to take rich, poor, black, white, red, yellow, and brown and make them a cohesive unit.

Coach Switzer ran into his bear in 1989 when he was edged out as coach at Oklahoma after a series of NCAA violations and serious misbehavior by a few players. There has been a lot of criticism of Barry for what went on. What people didn't understand was that Barry did what he did not because he didn't care but because he cared too much.

Barry Switzer has a heart that sees troubled kids not as what they are but as what they can be. He understands that the difficult life experiences of these kids leave them without the kind of values that make them strong, righteous men. Some lacked a father in the home or the proper discipline in their lives that all kids need. Because he believes young athletes could be more, he often went more than the extra mile to help, and in a few instances, it came back to haunt him. We're all human, even "the King of Oklahoma."

Sometimes I think Barry Switzer got scrutinized more than your average coach because he was anything but average. He didn't cater to the old guard in coaching who preferred the Woody Hayes or Bear Bryant model. He was only 37 when he became head coach, and his back-to-back national championships as a hotshot, confident rookie coach stuck in some people's craws.

He left OU the all-time winningest head football coach in the school's history with a remarkable .837 winning percentage, the fourth best percentage in NCAA Division 1 history. Besides the national championships, he led Sooner teams to 12 Big Eight Conference championships and 8 bowl game victories in thirteen appearances. As Ronald Reagan said when he left office, "Not bad. Not bad at all."

Eight years after his departure, Coach Switzer returned to OU in glory to attend the groundbreaking ceremony for no less than a sports complex being named in his honor. Time forgives almost everything, and Barry Switzer, who never shied away from controversy, deserved the recognition as one of OU's greatest coaches. At the groundbreaking ceremony, David Boren, the president of OU, joked, "It was first called the Barry Switzer Complex, but we wouldn't want anyone to think Barry had a complex, so it's the Barry Switzer Center."

Today, it's a magnificent addition to Oklahoma's extraordinary sports facilities. The center is home to the football offices as well as a locker room, equipment room, a 13,000-square-foot strength and conditioning complex, a sports medicine facility, and the Touchdown Club Legends Lobby.

Last year, Coach Switzer joined the other legends of his sport when he was inducted into the Football Hall of Fame. I am pleased to see a man who had done so much for me honored for what can only be called an extraordinary record. And part of that legacy is the friendship that Coach extended to his players. Barry Switzer told me once, "If you live to be 100 years old, you'll still only be able to count your real friends on the fingers of one hand." I've met a lot of folks over the years through football, church work, and politics. People of every political stripe often see me on television carrying the Republican Party's message instead of a football, so a lot of them feel they know me. They come up to shake hands or give me a piece of their mind everywhere I go, and that's great.

But when it comes down to my real friends, Coach was right; it's a small number that I can always count on to be there through the good times and the bad. When I was on *The Chris Rock Show*, he needled me in a good-natured way about being a Republican. He asked me if I hung out with a lot of "old white guys in three-piece suits." I laughed and answered, "I hang out with my family and my staff." Which was very true even if it's not the norm in Washington.

And real friends also tell you the truth when you need a little humility, as I did on several occasions my freshman year. There were many times when I walked off the field knowing that I hadn't played my best. But as I headed to the locker room, I could hear fans telling me what a great game I'd played, patting me on the back and congratulating me when I knew better.

Although people mean well, hollow flattery is also endemic to politics, and those who listen to its siren song often lose touch with the values and the vision that drew them to politics in the first place. It's easy to buy into that smooth talk. Who doesn't think praise is some of the sweetest music you can hear? But undeserved praise can wrap around your mind and soul like a tourniquet cutting off your own best judgment.

The media loves to build someone up just to tear them down. People get put on pedestals they don't deserve and sometimes don't want. Then they slip up, and the fall is usually a big one.

We live in a society that loves celebrity whether it's on a football field or a movie screen or Capitol Hill. But there is a mighty big difference between being a celebrity and being a hero. A celebrity is known for being known and maybe for a special skill or talent. Heroes are known for living their values, upholding their convictions, standing by their principles at all costs.

Heroes are those who survived the Great Depression like my parents or served in World War II. They are Vietnam War veterans, and Rosa Parks, Dr. King, and Jackie Robinson. Heroes are the people who put their lives and their safety on the line. The firefighters, police officers, and rescue workers who climbed the smoke-filled stairs of the burning World Trade Center to save others. The hundreds of fathers and mothers, sons and daughters, who were never seen again. And the members of the military serving us today in the battle against terrorism. Those are the real heroes.

I got my name in the paper a handful of times for scoring touchdowns at OU. That made me a minor celebrity, not a hero, and at OU, one of the most important parts of my education entailed learning not to confuse the two.

My time as Oklahoma's starting quarterback did give me confidence and enhanced my leadership abilities. Like my years at Eufaula High, it added another layer to the foundation on which I would build my life by teaching me that real leaders create an air of expectancy.

We must face each challenge expecting to do well, whether it's winning the big game as a quarterback or the battle for a lost child's soul as a minister or a hard-fought campaign of ideas as a politician. If we fail, we must learn why and never be brought down by the same mistake again.

I think all champions create that air of expectancy and confidence. Muhammad Ali did it. When he took to the ring, you expected him to do what he said he would do, and he rarely disappointed us. Roger Staubach had that air of expectancy about him, too. So did John Kennedy and Ronald Reagan. There are many other champions I could list who had the same intangible confidence that they would get the job done regardless of the odds or the obstacles. There's an old saying: "The person who believes he can do something is probably right, and so is the person who believes he can't."

I tried to exhibit a mental toughness while attending the University of Oklahoma that would give me both the ability and the authority to lead a championship team. Once again, I found that being a starting quarterback means doing more than calling plays or reading defenses. It means leading a team by creating that indefinable air of expectancy and confidence. I count myself incredibly lucky. At OU, I became something much more important than a football star or even a graduate. I became a man.

8

Climb the Mountain You Choose

A man is a lion for his own cause.

Scottish proverb

Graduation was only six weeks away when I went home one weekend to see Mama and Daddy. As we usually did, Daddy and I sat in
the living room after dinner solving all of America's problems according
to the Book of Watts. My father was deeply interested in public policy
issues—although he'd say that's a pretty fancy way of talking about people's troubles. We would often discuss everything from abortion to the
failure of the welfare system. Although Daddy was a lifelong Democrat,
he believed that abortion was wrong and welfare was a well-intentioned
system designed to deliver people out of poverty that more often
imprisoned them in a dead-end existence. He knew that some abused
the system because we saw it in our own neighborhood.

We'd talk about his business and my schooling; Frankie and the
kids; our faith and the future, his and mine. I was just about to go to bed
after offering up my "educated" view on the topic of the night, when
Daddy suddenly said very seriously, "You know, I think I'd like to go to
college."

This was a new one. Buddy Watts hadn't finished seventh grade, and now he wanted to go to college. I must have looked pretty incredulous when I asked him to explain himself.

"What do mean you want to go to college? You're 57 years old, and a double-bypass heart patient," I said. "You still work dawn to dusk with your cows and rental houses even though you shouldn't. You're pastor of a church. Why would you want to go to college?"

He said, "I'd like to see what makes you college guys fools after you get out. You all seem to lose your ability to use common sense."

As he often did, Buddy knew just when my ego needed a little deflating. He reminded me that education isn't always found in books. You can discover knowledge in all kinds of places but perhaps none more important than in your own heart.

Daddy's timing was perfect. Even though graduation was just around the corner, I still didn't know where life would take me next. My dream of playing pro ball was very much alive and well that spring. The New York Jets had scouted me in the Japan Senior Bowl, a postseason all-star game played in late January in Tokyo. That trip was a real eye opener for me because I saw an entirely different culture for the first time. I earned the Most Inspirational Player award in that game—not as quarterback, but as a member of special teams: returning kickoffs, playing wide receiver and defensive back. I was grateful for the honor. There's nothing a potential minister likes better than being called inspirational, but with the NFL on my mind, I figured the Samurai Warriors helmet that went with the award might come in handy.

New York Jets coach Walt Michaels joked once that "a man who has no fear belongs in a mental institution or on special teams." I didn't find special teams intimidating, but playing outside my usual quarterback position didn't let me showcase my leadership skills at a crucial time in the NFL scouting process. That was a disappointment.

Then some interest came my way from an unexpected source. I happened to run into Jimmy Rogers, an Oklahoma teammate who graduated two years before me. He had opted to sign with the Edmonton Eskimos, where he played alongside the great Warren Moon.

He urged me to consider the Canadian Football League.

"You'd fit in real well in Canada," he told me. "Your style is a lot like Warren's," meaning I could run or pass.

He turned out to be right. When my grandparents lived in Canada, I doubt if they ever imagined that one of their grandsons would also take up residence north of the border more than half a century later. I certainly never thought about it until I found myself sorting out my pro football prospects, and I didn't like my options. Canada came as a surprise, but it was a terrific turn of fate.

After a little investigation, I discovered that five out of the nine Canadian teams already had black quarterbacks. That told me the opportunity was there if I had the courage to go after it and the wherewithal to ignore the prevailing wisdom: nobody in their right mind turns down the NFL. Crazy or not, that's just what I did, but not because of an overwhelming interest in Canada.

Everything I knew about this big country to our north I'd learned the summer after seventh grade. Mama and Daddy packed up the family for a little vacation in Edmonton. We had a wonderful week. The culture was similar to ours but with enough differences that it made the visit a memorable one, especially for a family that didn't take vacations often. The Canadian money was confusing; but all in all, we thought Canada was pretty grand. I have to admit, however, I didn't give our northern neighbor much thought in the intervening years.

My agent, Abdul Jalil, did, however. The Canadian Football League's draft works a little differently than ours. The league has what's called a "negotiating list." Each club can put potential draft choices on the list, which makes the players off limits for their competitors. Suddenly, I found myself on the list as a possible choice of the Ottawa Roughriders. Only after the NFL draft is completed do the clubs get involved in serious negotiations with U.S. players.

The recruitment process, however, starts earlier. A couple of months before the NFL draft, Frankie and I found ourselves the guests of the Roughriders in the early days of March 1981. We both fell in love with Ottawa. Although it's the capital of Canada, with a population of 1 million people, it is one of the cleanest and friendliest cities we've ever visited. The beautiful Rideau Canal meanders through the city, which is marked by the imposing buildings of Parliament.

The nine Canadian cities that had football teams at the time represented about 80 percent of Canada's population. They were all as friendly as the people of Ottawa and had the same penchant for keeping their hometowns sparkling. We were impressed, but like most American players, I wanted to wait for the NFL draft before getting serious about a CFL career move.

The day of the NFL draft is a little like Election Day or the day your child is born. The anticipation nearly kills you. In my case, coming from a predominately running offense, I was going into the draft with a big disadvantage. The Wishbone is a tremendous college offense, but the NFL at the time generally favored the more traditional passing quarterbacks. NFL scouts look for guys who are used to dropping back into the pocket. Leading the Wishbone Offense, I ran more than I passed. In the NFL, there are only a few instances when quarterbacks generally run—quarterback sneaks for short yardage, bootlegs near the end zone, or a broken play when the quarterback has no choice but to go for yardage himself.

The 1981 draft was a two-day affair that began on April 28 at the Hotel Sheraton in New York. I was home in Norman with Frankie. I had decided that if I wasn't drafted in the first six rounds, I probably wasn't going to play pro ball, and for the most part, I'd come to terms with that scenario. Frankie and I spent the first day hanging around the house hoping the phone would ring. When a call didn't come, we both accepted it as part of God's plan for us. But the next day, God, apparently, had a change of heart. At the end of the seventh round, the L.A. Rams called. If I hadn't been selected by the tenth round, they told me, the Rams intended to draft me to play running back.

I wasn't thrilled with the idea of playing out of position, but the idea of living in L.A. was an exciting prospect that did have some interest for me. As it turned out, it was a choice I never had to make. In the eighth round, the Jets, who had shown some interest in me immediately after my senior season ended, took me as their second pick. I have always thought the Jets picked the two players with the most colorful names in the draft that year as their eighth round choices: Admiral Dewey Larry, Jr., and Julius Caesar Watts, Jr.

I had hoped to go higher in the draft, but I was happy to have finally made it to the NFL, or at least to a training camp. In the end, my NFL dream didn't happen, but I was the one who made the choice to go elsewhere.

In early May, I reported to the Jets' three-day rookie camp at New

York's Hofstra University. Like most rookies, I was excited and scared, hopeful and nervous. Once I was there, I asked the coaches what my chances of playing quarterback were.

"Slim and none," they told me. "We don't need help at quarterback. We need a possession back, a third-down guy who can return punts and also play wide receiver. What we're going to do with you will depend on how things fall into place during training camp."

Having no chance at quarterback was a real blow to me. Almost my entire high school and college career had been spent fine-tuning my skills at that position, learning how to lead a team. I had 8 years invested in being a quarterback, and now I get drafted to play out of position. All that effort was for naught. I also figured that 9 times out of 10 the first guy to get cut is the one trying to play in a new position. Most people who had seen me in action regularly thought I could throw the ball pretty well; we just didn't do it much at OU with Billy Sims, David Overstreet, Buster Rhymes, and Stanley Wilson in the backfield. That would have been a waste of good talent.

I tried to understand the Jets' point of view. Clearly, they had depth at the quarterback position, so their decision to put me in another spot made sense for them. Although I wanted the NFL desperately, I didn't want to find myself locked into a reserve role for my whole career. Quarterback was my identity, and they were suggesting that I jettison it like an old letter jacket.

I believed then, and do today, that more went into their decision than simply an oversupply of quarterback talent. During that time in the NFL, black quarterbacks were as rare as finding John Madden on an airplane. James Harris had done a good job in his backup role with the L.A. Rams. So had "Jefferson Street" Joe Gilliam with the Pittsburgh Steelers. But neither man was ever seriously considered as a starter.

I'm not sure I consciously understood the racial dynamics that were in play at the time. Even in the dark days of my freshman year when I thought about quitting, deep down I never believed I couldn't overcome any obstacle to become OU's starting quarterback if I got the opportunity. I just needed a little assurance from my family and coach that my chance would come.

But the NFL was different. There was no family or understanding coach to guide me. This was the big league. Millions of dollars are won or lost on the strength of the coaches' and scouts' decisions on talent. I don't believe any of the Jets coaching staff or management was inherently racist. It wasn't that they thought black athletes could not be good quarterbacks. But as I learned in high school and college, being a quarterback is more than having a particular set of physical or mental skills. It requires a certain leadership component, and that's where I believe the NFL met its bear.

Not unlike the rest of the country, NFL coaches and managers had a tough time seeing blacks in leadership positions, and they worried that fans would react negatively to putting blacks in charge of a team. By the late 1970s and early 1980s, blacks had made tremendous progress in all facets of American life. Sports was one of them, but hiring blacks as quarterback or as coach was still years away. I had finally run into a wall I couldn't scale: I wasn't just being impacted by race; for the first time in my life I was being defined by it.

That was the grim reality I faced the summer after graduation. I had to decide whether I was willing to give up my hard-won identity not only as a quarterback but also as a leader in order to reach my dream of playing in the NFL or choose another path that would allow me to remain true to my own self. With a wife and two children to support, the pressure was enormous.

Jalil urged me to head north—to sign with the Roughriders, prove myself as a quarterback, and then come to play in the NFL. I was willing to pay some dues, and I gave my agent the go-ahead to negotiate with the Roughriders and end talks with the Jets. A lot of people thought I'd lost all common sense, but I had my own reasons for choosing an unexpected direction—reasons that came from what I realize now was a growing sense of self.

My Canadian career almost didn't happen. Jalil and Ottawa's general manager, Jake Dunlap, were miles apart when it came to the contract terms. Neither one could seem to find a compromise we could all live with.

Weeks went by with no word of a deal. By then, I had written off the idea of playing any more football—tough for me to swallow, but with a young family, I needed to make plans for the future. I've always had an interest in the energy sector, and growing up in Oklahoma, you feel like you've got a little "black gold" in your veins even if you aren't lucky enough to find it in your backyard.

I wanted to work for an oil company to learn the business from the ground up, hoping to one day strike out on my own. I understood that business success, like winning football, requires study and solid preparation, and I looked forward to the challenge. Still, with the energy crisis of the mid-1970s still fresh in people's minds, it was common in those days to see would-be entrepreneurs put together limited partnerships to finance oil exploration projects. Thanks to my contacts at the university, I was able to meet with some influential Oklahoma supporters who were in a position to help me get a career in the energy field off the ground.

They quickly doused the idea with ice water. "This isn't something you want to do," I heard more than once. "Those investors are very sophisticated," all of them warned me. "They'd eat you alive."

I'd have been a tough meal, but I took them at their word and lowered my expectations. An entry-level position with a company was a safer road to travel. A few years earlier, they helped another quarterback, graduating with the same journalism degree I'd earned, get started in the energy field. We had both requested the same kind of backing. The only difference between us was the color of our skin. It made a guy wonder. Had the pool of investors changed or just the quarterbacks involved?

Or, like the NFL, was it too difficult for these people to imagine a black man capable of becoming a winner in this industry, capable of leading his own company? I'll never know for sure. But I do know some forms of racism are subtler than others, and it's a shame when lingering bias cheats this country of the potential of any person with a dream and the ambition to succeed.

Just when I had almost forgotten about football, one Saturday morning in late May, nearly three months after our visit to Canada, a surprise call came from my agent.

"We've got a deal with Ottawa!" he yelled.

At first I was speechless, which may have been a first for me. Finally, I managed to say, "Well, this is kind of unexpected."

But then, I got the Watts enthusiasm rolling. "Great! Tell them I'll get myself together and be there on Monday."

"You better get packing," he said. "You're booked on a flight to Ottawa leaving in four hours."

I madly threw some clothes in a couple of suitcases; kissed Frankie and the kids goodbye; and by 1 o'clock, I was on a plane headed for training camp in Peterborough, Ontario, a place I'd never even heard of. Once the initial rush wore off, I found myself wondering what in the world I was doing. I hadn't given the deal any real thought. Frankie and I hadn't made any plans about how or where we would live as a family. I hadn't even said goodbye properly.

Everything seemed to be happening too fast. I wasn't prepared, and I was getting a feeling that was uncomfortably familiar. Could this be my freshman year all over again, going to a strange place where I didn't know a soul? All I could do was put my trust in the Lord and my agent.

The Roughriders sent a driver to retrieve me and my bags from the airport in Toronto. I settled uneasily into the car for the two-hour trip to Peterborough. Most of the time I spent worrying. Seeing the beautiful countryside the next morning helped a little. I was to learn that Canada has some of the most spectacular scenery in the world—beautiful lakes and forests and truly breathtaking mountains. Of course, coming from Oklahoma, what I thought was a mountain would have been considered a rolling hill in most of Canada.

When we arrived at camp, I pitched into the practices, but over the span of a day and a half, my mood never recovered. All the warm feelings I'd had from our tour of Ottawa earlier in the year faded as I thought more and more about home. No matter how nice everyone was to me, I felt like a stranger in a strange land. At least I was honest enough with myself to know that I wasn't ready to compete for a position.

I went to the head coach, George Brancato, and said, "Coach, you don't owe me a signing bonus. You don't owe me anything. Just give me a plane ticket home and get me to the airport."

And just like that, I left camp. I went out to Oakland to spend a few days with my agent. I needed to talk things over with him.

"My career is finished," I told Jalil. "I just don't think I'm going to play football." Those few days were some of the most traumatic I've ever experienced, but in the end, I came to accept the idea that playing pro football just wasn't meant to be.

Back I went to Norman, mentally ready to resume my work at King Energy, an Oklahoma City oil company. It was time to start building that foundation for myself in business much as I had once laid the foundations for a career in football. I was just fine for about six weeks—no regrets, no withdrawal pains. Then, my buddies began reporting to NFL and CFL camps: Darrol Ray, Freddie Nixon, Richard Turner, John Goodman, Keith Gary, David Overstreet, and Billy Sims. And suddenly I realized that for the first time in a decade, I was not going to be playing football in the fall. I began to miss everything about football—the workouts, the strategy sessions, the scrimmages, even the sound of the coaches screaming in somebody's face. I wondered if I had made the right decision. But my pride kept me from calling and saying I wanted to come back.

A couple of months later, a call came from my agent.

"You're not going to believe this," Jalil said, "but Ottawa just asked me to see if you would reconsider and come back."

"Well, I think I probably would," I said with my heart pounding. No one wants to be a quitter, I thought to myself. I'll go back and see what happens, but at least I'll get it out of my system. The last thing I wanted was to spend the rest of my life wondering about what might have been.

Canada's pro football season begins six to eight weeks earlier than the NFL, so when I got there, the Roughriders had already played six games. They were just beginning a two-week bye, which gave me a chance to work with my new teammates and coaches and learn some of their system before the next game.

I spent the first few weeks in the Bytown Hotel before I finally found an apartment. Having spent most of the summer doing little to keep myself in shape, the first thing I needed to do was get some physical conditioning fast. I ran every day in the morning from the hotel to the stadium, worked out with weights, and felt my body come back fairly quickly.

Still, I sat on the bench for the next two games watching the team in action. Then, in game 9 of the 16-game season, I finally suited out for the first time as backup quarterback to Jordan Case, who replaced Condredge Holloway, traded to the Toronto Argonauts. The next two games Case played the first half, and I played the second. Over the last five games, I earned the position of starter. The team was 2 and 9 when I took over. We won three of the last five games to finish the season at 5 and 11.

At the time in Canada, there were four teams in the Eastern Division and five in the Western Division. The top four teams in each division go to the playoffs. So even with our less than stellar record and third-place finish, we advanced to the playoffs.

In the first game, we beat the Montreal Alouettes, who could boast Vince Ferragamo, Tom Kuzano, and my former Oklahoma teammates David Overstreet and Keith Gary. In the second game, we went to Hamilton and upset Frank Kush, Tom Clements, and the Tiger-Cats. It was a very tough game, and Hamilton led 13–10 in the final moments of the game. Then, my old Oklahoma training came in handy. Despite Hamilton's deep-zone coverage, I spotted receiver Pat Stoqua and moved to a pitch-and-run play—108 yards for the touchdown. We beat the Tiger-Cats with their 11–4–1 record by a score of 17–13.

Rufus Crawford, the Tiger-Cats halfback, talked about the loss after the game. "The worst part was that Pat Stoqua couldn't run a 7-second-flat in the 40-yard dash. But we couldn't catch him."

Another CFL player and former teammate at OU, Keith Gary, paid me a very fine compliment. He was a defensive lineman who had been drafted in the first round by the Pittsburgh Steelers, but had chosen to play for Montreal.

"You know," he told a reporter, "there's something about J. C. He always has a way of getting it done." If they put that on my tombstone, I'd die a happy man.

Our surprise win over Hamilton put us in the Grey Cup, Canada's Super Bowl, competing against Warren Moon and the mighty Edmonton Eskimos for the national championship. Many Americans think football is a fairly new sport to Canada. Not true. Canadian football has its origins in British rugby, and the Grey Cup dates back nearly 100 years. It's played in late November, earlier than our Super Bowl. Any later and the players and fans risk frostbite or worse from the deadly Canadian winters.

Conventional wisdom had it that the Roughriders didn't stand a chance against the three-time defending champion Eskimos, whose win-loss record of 14–1–1 dwarfed ours. We should have been intimidated, but I can honestly say we weren't. Maybe we were just too stupid to realize we were supposed to lose. My attitude was, "They're going to have to beat us to beat us."

The game got under way with Warren at the helm of his team and me at the helm of mine. It was a cold, overcast day. But soon the excitement of taking the lead got my blood running. Despite all the naysayers, all the sportswriters and critics who predicted a rout by halftime, the Ottawa Roughriders were ahead of Edmonton by an amazing score of 20–1!

Unfortunately, it wasn't to be a Cinderella story for the Roughriders. In the second half, our defense went south on us, and we sagged offensively, too. After a stormy third quarter for Edmonton, which saw Warren Moon replaced for a period of time, Warren and the Eskimos came back with a vengeance. Then a couple of critical calls went against us late in the game. One of them was a double interference on our all-star inside receiver, Tony Gabriel, a guy who had hands like glue. It came on a critical third-down play and took away what would have been a first down. Then, in a moment of exquisitely bad timing, I fumbled the ball inside our own 20-yard line. My heart sank as the ball slipped from my fingers, and nothing I or anybody else could do could get it back.

With 9 seconds left on the clock, Dave Cutler kicked a field goal and capped what one sportswriter called the "greatest comeback in Grey Cup history." Edmonton had edged us out, 26–23. While we lost what many fans considered the most exciting Grey Cup in many years, the fact that we nearly defeated the best team in Canadian football against all odds was considered by some a victory in and of itself. And I was again honored to be named offensive MVP despite our loss. But I would have traded twenty MVP awards to have won that day.

Coach Brancato later characterized the season this way: "That year I put together three teams. We had a bad team to start the season, then a fair one, and then a pretty good one."

It was great to get back home to Norman and the family for the off season. I really missed Frankie and the kids—even with Jerrell still in the terrible twos. There is nothing like reestablishing family connections to bring a guy back to earth, back to the roots that matter most. Mama and Daddy were both doing well and all my brothers and sisters were a welcome sight.

But before long, football was again on my mind for all the wrong reasons. After our near defeat of Edmonton, everyone had high hopes for the Roughriders for the coming season. Unfortunately, a contract dispute derailed my relationship with the team. My agent and I both understood that my salary was to be paid in American dollars. Management said no, the agreement had been for Canadian dollars.

This was more than just a fine point. Being paid in Canadian translated to about a 30 percent pay cut. Add the massive tax burden which is a fact of life in Canada, and I was taking a real hit in take-home income—and home was a long way off.

I decided to go to the CFL Players' Association for help. After all, isn't that what a union is supposed to do—protect the rights of its members from abuses by management? Apparently not, in this case. For me, that contract dispute provided real insight into labor-management relations. Neither side was on my side, and the Players' Association failed me miserably.

We seemed to have reverted to the negotiating stance that tied up the contract in the first place. I wouldn't budge and ended up sitting out the entire 1982 season. In 1983, Ottawa hired a new general manager, who viewed the world a little differently. He wanted me back, and ultimately, we struck an agreement for a four-year contract.

The first thing I did when I returned to the Riders was head to the team accountant's office.

"Carol," I said, "I don't want any money taken out of my check for the Players' Association." I didn't feel they had any claim to my loyalty or my money. They hadn't stepped up for me during my contract dispute and weren't very supportive during the year I sat out.

"Sorry," she said. "We can't do that."

"But I don't want to belong to the association."

"We still can't do that," she repeated.

"No, you don't understand," I replied. "I do not want to be in the Players' Association. Period."

"No, *you* don't understand. It's mandatory. You have to be part of the Players' Association in order to play in the CFL."

That was my introduction to the concept of union shops versus right-to-work. People wonder why I'm a Republican? My experience with a union that put its own power ahead of the welfare of its members taught me a hard lesson. Even more important, however, I saw firsthand what happens when the long arm of government extends into the delicate arena of labor-management relations. And I didn't like it.

On the bright side, I was back playing the game I loved in a city I was growing to love. Ottawa was a big enough city to be exciting, but still small enough for a country boy to be comfortable. Frankie and I worked out a routine to bring us together as a family as often as possible. I would usually leave in the middle of May for training camp each year. By the first week in June, school was out and Frankie and the kids would arrive with their suitcases stuffed with clothes and toys and everything else they needed to make Canada home away from home.

In late August, they would head back to Norman, and I would head onto the playing field. The season usually wrapped up by November 25. Thanksgiving was always touch-and-go, although a few years I managed to celebrate Thanksgiving in both countries. If the team had a "bye" weekend, I would sometimes scoot home just to recharge for a few days.

Much as I enjoyed playing in Canada, there were some drawbacks, too. As quarterback, you lead the team. The problem was knowing exactly who the team was. Turnover is a huge problem for the CFL. Once the NFL teams start to cut players, the team rosters in Canada seem in a state of almost constant change. We'd find ourselves in the eleventh game of the 16-game season, and we'd have to practically introduce ourselves to our teammates in the huddle. The Canadian talent was fairly stable, but the American players would come and go until October 1.

By mid-October, there was a cutoff point after which the players had to remain with the same team. For me personally, it was the turnover in coaches that caused problems. While I was playing for Ottawa, I had five different offensive coordinators over the course of four years. I learned something from each one, but it was disruptive for the growth and maturity of a young quarterback. I was relearning all the time, never staying with one system long enough for it to become as automatic for me and my teammates as the Wishbone Offense had been. You can't stop to think during a football play, not even for a nanosecond. Football must be instinctive. A quarterback must do exactly what's needed and at precisely the right moment.

That constant turnover was a real problem for the Canadian league and I suspect it still is. Successful sports teams often build their identity around certain outstanding players who establish themselves with the community. When I played Canadian football, U.S. players rarely stuck around long enough to create those connections with the public. It was unusual to see a player stay with a team for five or six years, and that has hurt the league over time. I'm not sure there's any easy answer. There are terrific American players let go by the NFL every year, and they add mightily to the quality of play, but I thought the long-term strength of the individual clubs was compromised to a degree.

When it came to competition, I found the atmosphere decidedly different in Canada, too. It wasn't that the teams and the fans didn't care about winning. They did, but not to the degree I was used to at OU. I always found their attitude a little more laid-back. They'd say, "Gosh, you fellas were sure entertaining tonight" instead of throwing chairs when things didn't go right. And unlike the United States, football wasn't the biggest sport going. Hockey was and still is. Boys in Canada didn't grow up wanting to be Warren Moon or J. C. Watts. They wanted to be Wayne Gretsky. You didn't see kids playing football on the church grounds or in somebody's backyard. You saw them in parking lots on roller skates playing hockey with a tennis ball. Where I grew up, we thought hockey was weird and football was king. Eventually, I came to understand and respect Canadians' love for hockey, and ironically, in Oklahoma City today, it's probably the hottest ticket in town.

Once the contract was finally settled, my first year back with the Roughriders was a good one. My backup quarterback was another Okie—Prince McJunkins from Muskogee, Oklahoma, a town just 35 miles from Eufaula. With a good wide receiver, Tyrone Gray, and a very fast halfback, Skip Walker, the Roughriders improved to 8–8 for the season, and I led the CFL with an average 17.7 yards per completion. But our playoff hopes were smashed by Hamilton, which was bent on revenge after the humiliating defeat a couple of years before.

The following year began even better, with the team off to a 3–0 start. In the fourth game, however, I got what's called "turf toe." It's a little like your garden-variety stubbed toe but much more painful and long-lasting. Even though it seems like a minor injury, it does seriously hinder mobility and hampered me for the rest of the season. I missed eight games outright, and I still feel the effects of my turf toe today.

In 1985, Kevin Gilbride came in as our new offensive coordinator. He was one of the best coaches I ever played for in Canada, and I would have liked to finish my career with him. By the final year of my contract, 1986, I was now in my late twenties, and I knew my time on the football field was limited. Sure enough, Ottawa brought in two younger quarterbacks that year. Then I got the order every football player dreads: "Coach wants to see you in his office." Five games into the season, the Riders had decided to let me go. At the time, I only needed one more game to be vested in the CFL pension system.

Luckily, three other teams lined up for my services, and I signed a two-year deal with the Toronto Argonauts. Toronto is only 300 miles east of Ottawa but is a much different city. With three times the population, it is a much more cosmopolitan city, with one of the most ethnically diverse populations in the world. Yet, like all of Canada, it retains a friendliness that belies its size.

I was happy to find a football home in this wonderful city for what turned out to be my last season in pro ball. Over the course of the last eleven games, I became the starter, and we made it to the Eastern Division finals before finally losing to my old rivals, the Tiger-Cats.

The door was open for me to stay in Canada a little longer, but I realized that my dream of going back to the United States as an NFL quarterback was over and my thirtieth birthday was just around the corner. Some people still tell me I should have stuck with the Jets, played special teams or running back, and counted myself lucky. But I've never regretted my decision to play in Canada.

I did put some feelers out to the NFL after the first couple of seasons with the Riders, still looking to play quarterback with a club that fit my style. Seattle and Washington seemed the most likely choices, but the interest wasn't there. The year I sat out, I also talked with the owner of the now defunct USFL New Jersey Generals, an Oklahoma oilman, but Ottawa had set the price to buy out my contract too high. So I happily returned to Canada to finish my career.

It was Warren Moon who finally made the move from Canada to the NFL as the first black Hall of Fame–caliber quarterback in the American league, and I believe he was the right man for the job. His talent was so phenomenal it simply couldn't be denied. Warren's statistics in Canada broke all records. He won the Grey Cup five out of the six years he was in Canada.

Warren was a free agent and could have signed with anyone. At 6 foot 3, he had the size and speed as well as the mobility of a great athlete. He chose Houston, and by choosing him, the owners made a statement about the role of blacks as leaders: "We'll just have to sell our fans on the importance of having this man on our team." This was a change in attitude, and Warren didn't disappoint.

In the style of Jackie Robinson, he always played well, with a quiet, confident demeanor that telegraphed he was no victim. There was no anger in Warren as he tore down the barriers and stereotypes that had kept black athletes from this coveted position. I believe the Lord chooses particular individuals to be the first for a reason, whether it is Warren Moon or Jackie Robinson or Sally Ride or Rosa Parks. God gives these special people the temperament and talent to effect his will.

I'm not saying football is a religious experience, although for some fans, I think it comes close. But I do believe God works in mysterious ways, and as I saw in high school, college, and the pros, sports has a proud history of taking us as a people beyond bigotry and hate.

Canada offered black quarterbacks the opportunity to succeed a little earlier than this country. There, people seemed to be much more open when it came to race. It's just not the issue it is here. One night when I was with the Riders, I went to a restaurant for dinner after practice. I was sitting by myself at one table; an elderly couple sat at another; and a few tables down, a group of six or seven college girls were having a noisy night out.

At one point, the elderly couple and I caught each other's eyes.

"They're having quite a party, eh?" one of them said to me, and I agreed. I don't know if they recognized me as the Roughriders quarterback, but we began to talk. Inside of 15 minutes, they'd invited me to their home for some wine and cheese. I declined, not being a drinking man, but I couldn't help but think that this would never happen in the United States. Few people—black or white—would invite a perfect stranger into their home. But that's how Canadians are.

I left Canada for home on December 7, 1986. I still wanted to win, still had most of the spirit it would take, but something was missing in my life, and I knew it. When I arrived home, I sat down with Frankie and told her, "I think God's allowed me to play my last football game."

I don't think she was enormously surprised, but she asked me what I wanted to do next.

"I don't know," I said honestly, "I'll just step out on faith and trust the Lord will speak to my heart."

Playing for the Canadian Football League didn't make a man rich in those days, and I had a family to take care of. So finding a job was my first priority. Thirty days later, I was the new youth director at Sunnylane Baptist Church in Del City, a suburb of Oklahoma City. The great basketball player Earl "The Pearl" Monroe said, "Sports is the only profession I know that when you retire, you have to go to work."

Youth minister as an occupation, however, wasn't high on my list. In fact, it wasn't on my list at all. But then, I didn't know John Lucas very well. Like a good football scout, John Lucas, the pastor of Sunnylane, had an uncanny ability to find the right talent for the right job. I'd met him a couple of times through mutual friends, but we were hardly close. Somehow, he sensed that I was the man for the job.

I had been home for a couple of weeks when John called and offered me the job. The idea of ministering was very attractive to me, as my faith had continued to grow during my years in Canada. But dealing with teenagers was not the ministering I had in mind.

"Oh, no," I thought as he told me about the youth ministry position, "Please, God. Not the youth. Give me the two-year-olds or seniors— anybody but teenagers." I remembered all too clearly my own youth and the buddies I ran with, my own mistakes and the toll they took. But John was not only a good salesman, he was a better friend who knew in his heart that this was the job I needed. So, with not just a little trepidation, I started what was to become an extended love affair with those young kids and that church.

Deep down inside, I wondered if the football bug would bite again in the spring when Canada's preseason rolled around in May. But by the time the Argonauts were scrimmaging in Toronto, I was already hooked on a new way of life.

I had learned a football team needs a quarterback to lead it, but I found out our kids do, too. After the "Dream Team" won the gold medal in basketball at the Barcelona Olympics, Magic Johnson said, "All kids need is a little help, a little hope, and somebody who believes in them." What a wonderful job description, and as a youth minister, the job was mine!

Some people didn't understand my decision to leave football for the youth ministry, but from almost the very first moment, this wonderful job taught me the power that each of us has to change the lives of others for the better. I found out that I had a lot to learn about dealing with kids, but I can honestly say that once inside the doors of Sunnylane Baptist, I never looked back.

9

In an Environment of Temptation, Kids Need Love and Authority

Children have more need of models than of critics.

Joseph Joubert

The giving of love is an education in itself.

Eleanor Roosevelt

I may never have looked back on my decision to leave the CFL, but that didn't mean I wasn't tempted once or twice to reconsider football—not to play again but to coach. I got to know Grant Taft, the head coach of Baylor at the time, through the Fellowship of Christian Athletes.

The FCA is a terrific organization dedicated to helping student athletes move toward spiritual growth by building relationships between family and church, and it has been a big part of my life since my freshman year in college. Today, over 500,000 kids participate.

Through FCA, I have met and made so many friends over the years who share a commitment to their faith and its ability to make our children all they can be. Grant Taft is one of them. In 1988, however, I have to admit Grant took me by surprise with a job offer that came out of nowhere. His quarterback coach was moving on—would I be interested in the job? This was a family decision. It meant moving to Texas away from family and friends, but more important, it meant leaving the job I had come to love and the kids who depended on me.

Frankie and I talked about what we knew was a wonderful offer. As far back as high school, I'd thought about coaching, and Baylor had been on my short list of places I would have loved to work. I knew Grant would be the perfect head coach to begin a coaching career under, and there was the salary, of course. Making ends meet on $300 a week, my Sunnylane salary, was often challenging. Using our savings from my years in Canada, we were able to get by but not much more than that. This job would be well paid. Still, I hesitated, and at first, I wasn't sure why.

At Sunnylane, Frankie and the kids and I were enjoying a normal family life for the first time since we had been married. Football had kept us apart, but here we were together and at peace. That stability and peace was hard to give up. I have always believed that the Lord guides us to take certain paths. Choosing OU; rejecting the NFL to play in Canada; forgoing the last year of my contract to return home—all were forks in the road. Yet I always felt that the hand of God steered me down the right path for my family and me.

This decision proved no different. I called Coach Taft and told him that I was staying put. "I feel God has plans for me here in Oklahoma." He was gracious about it, understanding my commitment to the kids of Sunnylane better than most coaches. As it happened, Grant Taft retired just two years later, so our time together would have been short-lived. But that's the name of the game in coaching.

There's a joke that's a staple of the coaching profession. A new coach comes in to take over for one that just got sacked. He sits down at the desk in his new office and finds that his predecessor has left him a stack of three envelopes. The one on top says, "Open me if you have a losing season."

As it turns out, the new coach's first season isn't so hot, so he opens the first envelope. "Blame it on me," it read. So he tells the sportswriters that next season will be better after the team learns his new system and forgets the old.

Unfortunately, the second season is no better, and he remembers the second envelope, which read, "Open me if you have two losing seasons." The note inside says, "Blame it on the alumni."

The coach goes out and tells the media that alumni pressures have been the problem, but things will take shape now that he's riding herd.

His third season is no better than the first two. He goes to the desk and finds the third envelope: "Open me if you have three losing seasons." The coach was desperate for some good advice and rips open the envelope. Inside is a note that says, "Prepare three envelopes."

The pressures and demands of successful coaching go far beyond just winning, as I observed at OU. Meeting expectations set by management, the media, the fans, or all three is the name of the game. Long hours, player problems, and a lot of time away from home go with the territory. At one point in my life, I would have been willing to embrace the good and the bad to spend a lifetime in a sport I loved, but by 1987 that moment had passed.

Muhammad Ali once said, "The man who views the world at 50 the same as he did at 20 has wasted 30 years of his life." Well, I was still a long way from 50, but my fascination with football as a career was over. That doesn't mean I don't still enjoy the game, and with a young son out there on the field, my interest in football has gotten a second wind. There's nothing I enjoy more than sitting in the stands cheering on Trey.

In the spring of 1987, Frankie and I settled down to a life of children, children, and more children. Not only had I taken on the 80-plus kids of the Sunnylane youth ministry, but the Watts family was growing, too. My daughter, Jennifer, had been born in 1984 when I was still playing in Canada. Julia, our last girl, joined the Watts clan in 1989, and I almost saw the delivery—almost. I could take an entire defensive line coming down on me, but I had never been very good at watching my wife give birth. Holding her hand in prelabor was one thing; the actual delivery was something else. After several hours of cajoling, the doctors and nurses finally had me talked into actually being in the delivery room with Frankie. I suited up in my hospital greens, but just as I was about to enter the delivery room, out came the doctor.

"Congratulations. It's a girl!" I had narrowly escaped again.

By the time Julia was born, Kesha, our first, was almost a teenager, and Jerrell wasn't far behind. For Frankie and me, parenting has always been a work in progress, as it is for most people. Both of us came from strong, family-centered homes. My parents weren't the touchy-feely types. Mama and Daddy didn't say "I love you," but we kids knew it just the same.

Frankie grew up in a tougher situation. Her mother was an amazingly courageous woman raising her children as a single mom. She did a tremendous job. Too often, I hear people assume a certain tone of moral indignation when talking about single-parent households, some in my own party; and I have to admit that kind of narrow attitude makes my blood go to a slow simmer. No one would wish a fatherless home on any child. I believe in my soul that the traditional two-parent—mom and dad—home situation is far and away the best for any child.

But sometimes, that's just not possible. And it's a testament to the hundreds of thousands of single parents, like Frankie's mother, who raise good, God-fearing children under the most difficult circumstances—children who grow up to become successful men and women in their own right. Like many children raised without a father in the house, Frankie struggled with that loss until early in her adult life. It even caused some challenges in our marriage. Frankie often shares that struggle with church and women's groups. Being open and candid about her difficult younger years has helped her and our relationship with each other and our own children.

Both of us *are* the touchy-feely types with our kids. We tell them all the time how much we love them. Nothing compares to the feeling you get when you put your arm around your children, telling them they are safe and loved. When we had Kesha and Jerrell, however, Frankie and I weren't much more than kids ourselves with a lot to learn. I sure didn't know much about being a parent. At first, I just followed Daddy's lead. I raised my children as I had seen my own father raise his. Anytime a difficult situation came up, I just did what Buddy Watts would have done. I think that's how most people get started in the parenting business.

Sometimes, I think Kesha and Jerrell were the guinea pigs of the group. They had to live through the worst of their parents' mistakes, but they survived us. In time, I understood the kind of upbringing that had worked on me, a slightly wild-eyed, mischievous kid, might not be right for my children. Daddy parented boys and girls the same, but I have found myself parenting my two sons with my head and my three daughters with my heart. That doesn't mean I don't love them all the same, but, at the risk of sounding sexist, I do believe that there is a difference between raising boys and girls.

All of our children have very different personalities, and we treasure each of them. But we live by one set of basic rules in the Watts household, and we have found that encouraging our children to obey those rules requires different approaches to match their individual natures. As I've grown as a parent, I've learned that one of my most important responsibilities is getting to know my children better—their likes and dislikes, their fears and dreams. Paying attention to our children, listening to them, is a big part of being a parent that I take seriously.

When it comes to knowing our kids, however, Frankie is the MVP. She is amazing, putting in 14-hour days, carpooling kids, packing lunches, helping with homework, and keeping up with housework. She's even managed to make three Little League games in one night. Believe me, I thank God for her love and commitment every day. I depend on her judgment and look to her wisdom when major family decisions need to be made. In this case, Mother knows best. She's with the kids every day, and nothing gives her more credibility in my book than her work, and I do mean *work*, as a committed mother. Mothers epitomize what a real hero is.

I think both of us wish we'd had the experience of child rearing we have now when we embraced the Sunnylane youth ministry. Instead, we took on this great new adventure with little more than faith to get us by. The glitz of pro ball was replaced by coupon clipping, endless scrimping, and what often seemed like round-the-clock ministering to our four children and a flock of emotional teenagers riding the usual roller coaster of puberty.

Sunnylane was a medium-sized church on the outskirts of Oklahoma City with a mostly white, blue-collar congregation. A black, ex-Canadian pro football player was probably not what most of the members expected as their youth minister. John Lucas had to do some convincing—first with me and then with his congregation—but he got his way on both counts. Years later I learned that a few people actually left the church over my hiring. John kept it from me at the time.

When I became youth minister, I wasn't altogether sure what I had agreed to. John, as he always could, helped me find my way. "J. C.," he said, "we're not trying to make kids Bible scholars here. Sure, there's room for that, and we don't want to ignore their Biblical instruction. But our number one priority is to love these kids, and let these kids know they are loved." Such a simple but powerful concept.

So that was it. That was my job. Love these kids, be their friend, and nurture them in the gospel. I had never had any formal training, but I did have a pastor who was willing to take a chance on me.

My first few weeks were a testing period for all of us. I had close to eighty kids ranging from grade 7 to grade 12. I found myself dealing with everything from trying to convince a 13-year-old that his voice really would change to counseling kids facing the typical peer pressures of sex and school. While luckily we didn't have serious drug problems to deal with, we had just about everything else. With few exceptions, the problems these kids brought to me were things I could relate to because I had experienced them at some point in my own life: pregnancy, relationships, problems with parents, trouble at school.

I spent a lot of time with "my kids"—usually meeting together at least three times a week. They were all as different as their problems. They were from single-parent homes, from two-parent homes, black kids, white kids, Latino, Native American—they were a reflection of Oklahoma and, in fact, the country.

Often, they would bring friends or visiting cousins to our meetings, and I didn't know from week to week how large the group would be. But from my point of view, the more kids gathering at Sunnylane the better. When it came to rustling up enough food or finding a few extra chairs, we always managed. It wasn't exactly on the scale of loaves and fishes, but we usually could accommodate whoever showed up for one of our gatherings.

One of our most popular events was called the Fifth Quarter. After home football games at Del City High School, we'd invite the kids over to the church for food and fun until midnight. It gave them something positive to do and kept them off the streets where they might get into trouble. My own high school days had taught me that teenagers can often take a wrong turn when they're left to their own devices. At Sunnylane, we wanted to give our kids an alternative.

Over the eight years that I spent in the youth ministry, I learned a tremendous amount about life, our children, and our responsibilities as parents and members of society. For me, it was fascinating to be on the inside of so many parent-child relationships. I was hardly an expert myself, but I took the pastor's mandate to heart. I was able to form really deep friendships with many of these kids. They learned that regardless of the problem, I wasn't there to judge but to help them find the right path out of their difficulty whatever it might be. Society has enough judges. I was never looking to see through a teenager, but instead, I wanted to see them through whatever concern or circumstances they faced.

Perhaps that's why we had a close-knit youth group. I came to believe that a successful youth ministry is about trust. Teenagers are pretty good at reading people. They know a phony a mile a way. When I told them, "We're in this thing together. I'll stick with you," I meant it. And they instinctively knew I meant it.

I found there was no difficulty they came to me with that I could not love them through. It was enormously fulfilling for me. Teenagers are always going to be teenagers and do teenage things. We can't protect our kids from every mistake they're likely to make. They're rookies in the game of life, trying to break into the big league of adulthood. They can and do make bad choices. Even though the youth staff and I tried to be their friends, nourish them with the Gospel, which is still the best parenting manual around, and encourage them to lead their lives in a Christian way, they faced temptations and didn't always defeat them.

For today's teenagers, this is the best of times and the worst of times. The possibilities open to them are limitless. Technology has created a whole new world of dreams for young people, but it has made this the most tempted generation in humankind's history. If the Internet is a wonderful window to the world, it can also be a pathway to pornography. Video games can be as innocent as a game of tag or they can mimic the false thrill of violence and aggression. Television and movies can educate our children about life, but they can also teach them about sex long before they understand the value of love and commitment.

In this confusing, complicated environment, kids need love *and* authority. As we grow older, we have the experience to understand shades of gray and moral dilemmas. Kids, however, need to understand right from wrong in an unambiguous ethical environment. This was the first, and probably the most important, lesson I learned as a youth minister. I also learned that it's better to start out tough and soften up, not the other way around. That was something I knew from playing under so many great coaches over the years whose jobs in some ways weren't all that different from mine as a youth minister. We were all trying to get a group of talented kids to make the right choices and be the best they could be.

From day one, I understood that part of being an adult friend to a teenager means also being a disciplinarian. One of my first activities as youth minister was a hayride for the kids. We were having a big time, singing and laughing under a big, black Oklahoma sky as we rolled at a snail's pace behind an old tractor. One of the older kids, along for the ride, was a sort of leader in the group who let his ability to sway the rest of the kids get the better of him at times. In fact, it often seemed he led his buddies in the wrong direction almost as often as he did the right one. Sometimes, he could be a bad influence.

His behavior was more rowdy than serious, silly shoving and pushing, and seeing just how far he could push me. It was the thrill of getting others to follow that seemed to motivate this kid, but for the well-being of the group, that kind of behavior simply couldn't be tolerated. I put my foot down. We had words, and later I had words with his parents. That was all it took. After that, it was smooth sailing. Some of the sponsors and even some of the other kids came to me amazed. "We've never had a youth minister who did that to him," they told me.

I'm proud to say my first "problem kid" has turned out to be a great success. He is married with kids of his own and doing very well. I've seen him a number of times since those days. Years later, I was touched when he thanked me for being an authority in his life. It reminded me of how I, too, came to understand and appreciate all those times Daddy yanked my chain to keep me headed in the right direction.

The youth ministry helped teach kids right from wrong, but my job was to also help them deal with those times in life when despair and disappointment come, as they inevitably do to all of us. Meredith Dan was one of the most popular girls in the youth group, always cheerful, with a beautiful smile on her sweet face framed by long dark hair. In the blink of an eye, this young girl found herself in a hospital room fighting for her life after a terrible car accident. One Saturday night, about twenty of us piled into cars and headed to the hospital after a pizza party. Everyone was in great spirits, including Meredith, whose release was scheduled for the coming Tuesday. We talked about what we'd do after she got out. She was so excited and happy that evening to be going home soon. The next morning, my phone rang and I learned she had died during the night. A couple of years after Meredith's death, another boy in our group was killed in a motorcycle accident.

Teenagers are so full of life. They're at a stage where their size and strength have just increased almost exponentially, their sexual vibrancy has come on line, they're nearly adults, and they feel invincible. Death is almost incomprehensible to them. "That's what happens to 'geezers,' not 16-year-olds," they think. Dealing with the deaths of two friends was a painful process for our kids and for me, too. I tried to put it in perspective, to give them a rock to hold on to in the storm of emotions that swept over them.

But even as they were recovering from their loss, we experienced another death that was even more startling. John Lucas, our pastor and the man who had guided me on my journey as a youth minister, unexpectedly died during surgery. He was 52 but really a kid at heart. It happened while we were at church camp, one of John's favorite pastoral activities. One of the most difficult moments in my life was the walk back to the cabin to tell the kids that he was gone. Nothing prepares one for times like these.

We all cried together that day, the kids, the sponsors, and even the cooks. Once again, the kids were confused by the Lord's decision to take someone who meant so much to so many. I told them that death is often a mystery that only God solves for us in time and that our prayer sometimes has to be "God, I can't see your hand, but I trust your heart." I reminded them of the Bible's words: For every thing, there is a season, a time to live and a time to die. And I told them that they must take heart in knowing that each of us would one day be called home to the Lord as John had been; that we would see John again, if we knew Christ as our personal savior. In many ways, as I tried to comfort these children in a

time of grief and confusion, I also found that my own faith strengthened. Pain tested our beliefs.

In some ways, I think they just wanted someone to hurt with them.
In grief, words provide little comfort. All you can do is put your arms
around them and cry with them. I did a lot of that with the kids in my
ministry. I said, "I know you're looking for some magical phrase, some
words that I don't have to make everything better. I just want you to
know that I hear what you're saying, and I'm here for you. We're going
to get through this thing together."

If this was a growing time for my kids, it was much the same for me.
Just as I had learned to lead a football team, I came to understand that a
youth ministry is a team effort, too. About three times a month, I would
schedule an activity for the kids on weekends; and every week, I ran
Sunnylane's Youth Sunday School. With eighty kids and three to four
events a month to plan and supervise, it was clear that this was no one-
person job. I had a volunteer staff of about nine people who helped me
and who were also a source of inspiration. I'd always been the kind of
person who said, "I can do it myself!" but an effective youth minister has
to rely on volunteers. Like most people who volunteer, these folks didn't
expect anything in return, which taught me the value of the words
"thank you." I owe many thanks to the volunteers and parents who loved
our Lord and our kids enough to get involved—volunteers like Nelda
Valdez, a single mom, who was extremely loyal to the Youth Ministry
and an inspiration to us all.

Some of my most precious memories are those years I spent working with the volunteers as we tried to figure out how to help our young
people grow. How could we do it in a fun way that would effectively
challenge them spiritually? Together, we had the privilege of seeing the
light of understanding begin to burn in the eyes of most of these kids.
They began to see that there really is more to life than its material side;
that life has a greater purpose than just having a girlfriend or boyfriend,
a hot car, a crazy haircut, or cool clothes.

They began to understand that the Lord quietly speaks to your heart,
and when you ignore your spiritual conscience, you are likely to make a
wrong turn. I hadn't always understood that God gives each of us the
measure of faith, and it's up to us to mold it and strengthen it by using it.

As someone who spent most of his life training his body to meet challenges, I found as a youth minister that faith is much like a muscle. The more you exercise it, the stronger it becomes. And as the years passed, both Frankie and I came to believe that in those last days of the 1986 season, it was God who brought me to a crossroads in my life to pause and reflect on what had been and what was to come. It was a time for me to "walk by faith and not by sight."

When I remember that changing point in my life, I always think of the scene when Indiana Jones, in search of the Holy Grail, is stranded on the edge of a deep chasm. Summoning every ounce of his faith, he steps out into thin air as stones magically appear one by one to give him safe passage. That's a little like how I felt the day I said yes to John Lucas and his faith in my ability to do the job. Thanks to the good Lord, Brother John, and the many Sunnylane volunteers over the years, the stepping-stones were always there for me, too.

10

Don't Take Everything You See or Hear in the Media as the Gospel Truth

> I consider the most enviable of titles the character of an honest man.
>
> George Washington

If Buddy Watts and Lucious Selmon steered the direction of my early life, Don Nickles gets at least some of the credit for one of the most important decisions I would make much later in my life—the decision to become a Republican.

It was spring of my junior year at OU, and I was studying for my degree in journalism. As a class assignment, I was sent to cover a U.S. Senate campaign debate between the young Republican candidate, a businessman from Ponca City by the name of Don Nickles, and the Democrat mayor of Oklahoma City, Andy Coats. They were both articulate, charismatic men who impressed me with their sincerity and intelligence, but I left the hall that afternoon one confused African American because I found myself agreeing more with what the Republican had to say than the Democrat.

Like most African Americans, I was raised in the most partisan of households—100 percent Democrat. Daddy was a Democrat. Uncle Wade was a Democrat. In fact, every black I knew was a Democrat. It never crossed my mind that a black person could actually *be* a Republican. African Americans are much like loyal union members when it comes to politics—expected to carry the flag for the Democrat Party, no questions asked, and more often than not, they do. I was doing no more than following tradition.

When it came to politics, my education was woefully thin. Most of my time in college was spent poring over playbooks, not partisan policy papers. While I called myself a Democrat, I couldn't have given you five reasons why—just one: blacks were supposed to be Democrats.

So when I arrived at the Nickles-Coats political debate, I assumed from the get-go that I would prefer the Democrat candidate. How could it be otherwise?

But as I sat there listening to the two men debate a wide range of issues from agriculture to government spending, I found Don Nickles's views resonating over and over again with me. He talked about the same values that Daddy did—faith, family, hard work, the importance of small businesses, and taking personal responsibility for your actions.

"Wait a minute," I thought, thoroughly confused. "This can't be a Republican—and a white Republican to boot—agreeing with my daddy." But that seemed to be the case, and I left that afternoon with a small seed of doubt planted by the persuasive arguments of the Republican businessman who would go on to win the Senate seat in November.

Nine years later, that seed which had put down tentative roots as I worked in my youth ministry and tried my entrepreneurial wings in business, finally broke ground in 1989. The year before, I had voted for Democrat Michael Dukakis, just as I was expected to. But in my heart, I knew I had compromised what I truly believed, and I swore to myself that it was for the last time. I switched my party registration to Republican, and to my surprise discovered that I didn't grow horns and a tail overnight.

That was the easy part. Breaking the news to Buddy that I'd defected to the enemy was the bigger problem. I was probably more afraid of confronting him than any lineman I ever faced in college or pro ball. Much to my disbelief, Daddy took it fairly well. He understood the substantive reasons for my decision to switch. He even agreed with most of them, but Daddy was an old-school Democrat. The ties that bind African Americans to the Democrat Party are strong and, I would argue, are often based more on geography or tradition than ideology.

A reporter from the *Tulsa World* once asked Buddy, who loved a good joke, about our political differences. "I'm not like my boy," this dyed-in-the-wool black Democrat told the reporter. "I told him that voting for the Republican ticket made as much sense as a chicken voting for Colonel Sanders." Since then I've probably had twenty reporters ask me about my daddy's partisan viewpoint. What they don't understand is Buddy Watts's sense of humor. He loved to get a rise out of people, especially "city slickers" with tape recorders and notebooks who always underestimated ol' Buddy Watts. He never wavered in his Democrat registration, and I respected his choice, but he also voted for more than one Republican presidential candidate in his lifetime. And he never voted for Bill Clinton. In the end, he came to respect my choice as well.

My uncle Wade took it a little harder than my father. In his mind, the Republican Party worked against the interests of "poor people, working people, and common people." I totally disagreed with his point of view and told him so. Later when I ran for political office as a Republican, he told Aunt Betty, "Now the Negroes won't vote for him because he's a Republican, and the whites won't vote for him because he's a Negro."

Thankfully, as it turned out, Uncle Wade was wrong on both counts, but based on black voting patterns then and now, his prediction wasn't particularly farfetched.

In a 2000 research study, the Joint Center for Political and Economic Studies has found that 74 percent of African Americans are self-identified Democrats; 20 percent are independents, and just 4 percent call themselves Republican. But younger African Americans are moving away from both parties. In the 18- to 25-year-old age group, 36 percent identify themselves as independent.

The study found a number of interesting facts. When asked which party has the better approach to dealing with a variety of issues, African Americans overwhelmingly chose Democrats, as I once did. But 74 percent of African American households with children support parental choice in education, one of the centerpieces of the Republican education reform agenda and anathema to the Democrat Party and the teachers' unions. Fully 76 percent of blacks favored Democrats over Republicans on Social Security issues, but 45 percent supported President Bush's Social Security proposal to allow personal investment accounts, while 42 percent opposed them, an admittedly narrow margin but an ideologically and demographically significant one.

As more and more African Americans have moved into the middle class, they are plagued with exactly the same problems everybody else in the United States faces—taxes, education, health care, Social Security, crime. Historically, black people have been conservative in their values, which revolve around family, church, and community. I don't believe most African Americans deep down share the Democrat Party's liberal cultural values. It was only when they vested their beliefs in the Democrat Party and allowed government to control their lives that they encountered deepening poverty, decaying families, and a sick welfare system that penalized women for saving money and for wanting to marry the father of their child.

I saw it growing up in Eufaula. It was a lesson lived in my neighborhood every day of the week. Yet ties are strong to the Democrat Party and memories long when it comes to the insensitivity the Republican Party has often shown on issues of civil rights, poverty, and racism over the past 40 years. Ironically, the party of Lincoln has, in many ways, forgotten its roots, which explains the near total control of the black community by the Democrat Party.

Jack Kemp rightly said that it would be a tragic mistake for the party of Lincoln and Douglass to concede the support of minority Americans to the Democrats. It would be a betrayal of the party's history. And the Democrat Party has its own history to deal with, too. I once did an online chat for the *Washington Post,* and someone e-mailed me this question: "How can you be in a party with a guy like Strom Thurmond when he said in 1948 'We don't have enough men in the Army to desegregate public schools'?"

I wrote back, "You can look at the early 1960s when we had a Democrat who stood in the door of the University of Alabama and said, 'As long as I am governor of this state, no Negro will walk through these doors.' Or go back to the same time and see a Democrat National Committeeman by the name of Bull Connor who turned fire hoses on black people because they wanted to sit at lunch counters in Birmingham. Or today, we've got a U.S. senator on the Democrat side who used to be a member of the Ku Klux Klan." There is plenty of shame to go around in both parties.

My decision to become a Republican wasn't based on history but on the principles that the party stands for today. By 1989, I felt I no longer had much in common with the national Democrat Party, which was out of step with who I had become through the experiences you have read about on these pages.

In December of that year, I switched my registration not out of convenience but out of conviction. I knew what I was doing wouldn't be popular. It created some strain, even in relationships I had built over the years. But I knew in my heart that this was the right road, the honest road for me to take and remain true to my own principles.

I didn't try to convince Frankie, though. And she didn't change her registration until some time later. Once I'd changed mine, however, I found myself more interested in becoming active in politics. My "coming out," so to speak, occurred one morning when I stuck a Republican campaign sign in my front lawn. I was mighty proud of that sign and took a second look as I headed out for a meeting in Houston that day. Frankie's reaction was a little different. She thought some local Republican hooligans had put the offending sign in our yard, and she wasn't having any of that. She marched outside, grabbed the sign, and threw it in the garage. We laugh about it now, but I don't think she saw much humor in it at the time.

Becoming a Republican was a process of evolution for me. Hearing Don Nickles played a big role, but so did a number of my own life experiences. One of those was my foray into the world of business. In the late 1970s and early 1980s, oil was going for $30 a barrel, which was good for making money in the oil business but the root cause of the country's economic problems. With high interest rates, high inflation, and long lines at the pump, something had to give, and it did—the price of oil.

They say timing is everything, and I believe it. Unfortunately, I managed to pick one of the worst possible times in recent memory to invest in oil exploration just as the oil and gas industry plunged into a down cycle. There was nobody in the state of Oklahoma associated with the energy industry who didn't get bloodied financially during these years. I saw good friends go through divorces and bankruptcies. It was a tough time for all of us.

Looking back, I probably should have declared bankruptcy myself, but whenever I considered it, I would hear Buddy Watts telling me, "Boy, if someone extends you credit, it's your responsibility to do what you have to do to get the debt satisfied." Those words meant something to me, and I decided not to take the easy way out. I've satisfied those debts and did what I thought was the honorable thing. It's never easy paying for mistakes, but at least I can sleep at night. Personal responsibility is an integral part of the Republican philosophy. My investing experience, difficult as it was, showed me the value and truth of that philosophy.

Becoming Sunnylane's youth minister didn't help our financial footing. Much as I loved my youth ministry, my salary simply didn't cover the bills of a growing family. As our savings from my football days dwindled, we cut back, scrimped, and saved where we could. We did what Buddy and Helen Watts and Frankie's mother would have done. But I found myself tossing and turning at night, worrying about how we were going to pay the bills and keep the kids in clothes and food from week to week. Finally, we admitted to each other that no matter how we tried to stretch the family budget, we just couldn't make ends meet on a youth minister's salary.

I knew the church couldn't afford to pay me more, so I decided to test my entrepreneurial abilities while continuing to serve the ministry. I accepted some paid speaking engagements and started a highway construction company, Ironhead Construction—those college summers working on road crews paid off. We did concrete milling, preparing the roadbed for the paving companies that would come behind us. I also began Watts Energy, a company to market fuel to military bases.

Beyond gaining a better appreciation for those willing to take risks in this life, I also learned firsthand how overregulation stifles the entrepreneurial spirit in America today. Although most of the regulations affecting business were created with the best of intentions—to protect consumers from unsafe products, to protect workers from unsafe workplaces, to ensure business pays its fair share of the nation's social costs, to protect the environment—the truth is, we've made it crushingly difficult to start and keep a small business going. Small businessmen and women often face as much regulation and licensing, insurance, and tax obligations as a Fortune 500 corporation.

Meeting both a payroll and the regulatory burden put on small businesses by government can often be mutually exclusive—especially for a new business. As I tried to build my highway construction business and my energy company, my own frustrations grew, and I was soon rethinking the role of government in our lives from a whole new perspective. Now, I knew from personal experience that government red tape can put many small businesses in a regulatory chokehold that can be fatal.

I also discovered I wasn't alone in my thinking. There were plenty of small business owners just like me experiencing the same kind of problems I was. It was just about that time that I got to know Ray Freeman, a black man working in the State Highway Department in Tulsa who also happened to be a Republican. Ray is a great listener, and it wasn't long before we became friends.

One afternoon, we were having lunch, and I was making my usual round of complaints about government regulation when I jokingly suggested, "Maybe I ought to run for office." After spending the last 10 years getting to know a lot of different politicians from both parties, I've found that, like me, many of them first got involved because of frustrations they had personally experienced with government. In those days, however, it was the politicians who were always after me to sign fundraising letters for them or to use my name for one purpose or another.

"Maybe I could run to change the system from the inside out, rather than the outside in," I told Ray over a full plate.

"Man," he said, "I think you could do it, and I know just the guy you need to talk to: Tom Cole."

I began to think seriously about what had started out as nothing more than an offhand comment. I had never had a burning desire to run for political office like some. Newt Gingrich and Bill Clinton both planned their rise to power from their high school days. I had my eye on a football career. Still, I thought, some of the other skills God had given me might provide me with some of the credentials for a political career.

I was never shy, I liked people, I was interested in policy issues and enjoyed public speaking.

In his book, *Bootlegger's Boy*, Barry Switzer tells a story about me from our trip to the 1980 Orange Bowl. I gave a talk at the University Baptist Church in Miami on the Sunday before the game, and Bobby Bowden, the legendary coach of our opponent, happened to be in the assembly. Just before the kickoff, he came up to Coach Switzer and said, "Barry, I had a chance to hear J. C. Watts speak. I went back and told my coaching staff, 'I'm gonna warn you, if that kid can play football as well as he can speak, we are in for one heck of a night.'"

When your father and both uncles are all known for their speaking, it should come as no surprise when the next generation exhibits some oratorical ability. It comes naturally in the Watts family. Over the years, I had become a fairly good speaker, preaching in the youth ministry and appearing at sports banquets and other local community events. I found I enjoyed it—enjoyed meeting people and discussing everything from teenage driving to the price of oil. Politics was a natural extension of the community interaction that filled much of my life during my years at OU and at Sunnylane. But a political career was never my driving ambition. In fact, it still isn't today. I never intended to be a politician forever, but just as the Lord guided me north to Canada and then back home again to Sunnylane, in that winter of 1990, it was his hand leading me down yet another path, and this one led to the door of Tom Cole.

Tom was then a state senator as well as a political consultant. What I knew about running for political office might have filled a short paragraph. I needed professional help, and according to Ray Freeman, Tom was just the man for the job. I had already given some thought to the spot I might go after—a seat on what's called the Oklahoma Corporation Commission—a nondescript name for one of the most powerful regulatory bodies in Oklahoma. The OCC oversees gas, oil, public utilities, and trucking, and its decisions have broad implications for the financial well-being of businesses and consumers alike.

Given my longtime interest in energy issues (I'm still fascinated by the oil and gas business today), I saw this as an opportunity to positively affect energy policies, to ease the regulatory burden particularly on small start-up businesses as well as to encourage new economic growth in our state through attractive regulatory policies. I knew that without affordable, abundant, and reliable energy supplies, our economic engine would slow and eventually halt. Both business owners and consumers would pay a steep price for that scenario. Once again, I was breaking new ground in seeking the seat. No African American had ever been elected to serve on this prestigious and powerful commission, but then no football player had either. To make matters worse, only two Republicans since statehood had been elected to the OCC. From my point of view, there was no black or white view on energy issues any more than on education or Social Security. After my experience with the NFL, I didn't intend to let race define me as a candidate, and I approached Tom Cole with that mind-set.

We connected almost at once. Tom is a warm, friendly man with a sincerity and a sense of integrity about him that instills trust. In some ways, he reminded me of my coaches. I had had complete faith and trust in both Paul Bell and Barry Switzer, and they never disappointed me.

Tom seemed to approach campaigning much as my coaches had prepped for a big game. He never underrates the opposition. He understands the importance of a well-conceived game plan with plenty of options ready for every contingency—kind of a political Wishbone Offense. It was a comfortable fit for both of us.

He's also a walking encyclopedia of political information and he's enormously fair-minded and articulate. There's no question in my mind that I was delivered into the right pair of hands, and my first political victory owes plenty to the intelligence and instincts of Tom Cole.

Early on, one of Tom's partners, Debbie Snodgrass, asked me if I was a liberal, a moderate, or a conservative Republican. I had barely come to terms with the fact that I *was* a Republican, much less tried to dissect my new party allegiance.

I had to admit I really didn't know the answer to her question. What I did know was that government was too big, too intrusive, and often made life harder for people, not easier. My first test was a controversial issue then pending before the OCC. The debate was over an unexpected windfall profit that the phone company had made. The question at issue was what to do with the excess profits. Do we allow the company to keep their profits and invest in infrastructure, which is a legitimate position, or do we allow consumers to keep a little more of their money to buy school clothes or put food on the table?

I saw it as an issue of fairness—what's a fair rate of return on equity for the company that still ensures consumers get good service, a quality product, *and* a fair price, too? You walk a fine line in regulatory actions wanting to encourage and reward good management but at the same time, whenever possible, give a break to those who've paid the freight. It was a tough call, but I came down in this instance on the side of the ratepayers.

President George W. Bush's budget is a perfect example of a similar balancing act. Winning America's new war on terrorism demands that our military forces have the best weapons and technology possible, and they don't come cheap. We must provide the funds to protect people here at home, too, through increased homeland security efforts. Short-falls in the educational performance of our children also demand increased support, and Social Security and Medicare must be protected. But with the country still in economic recovery, many of us also believe that the best prescription for a sick economy is a good dose of tax relief.

Others want to keep that money in Washington to create new government programs and more spending that will only make the federal government balloon in size. The Congress and the president must negotiate the right balance of spending that will help us win the war, increase homeland security, meet our social needs, and create jobs by sending money back home to people who need it and will spend it on family not government needs.

I've found in my short time in Washington that when it comes to spending other people's money, too many politicians forget how real people live. Real people clip grocery store coupons every week, trying to save 40 cents on a box of cereal or 25 cents on a gallon of milk. Businesses look for ways to cut costs. Single moms do it, single dads do it, and families do it. They save money.

When I was a kid, my parents pinched more pennies than Scrooge. Coach Switzer, who got to know my dad pretty well, once joked, "I think Buddy Watts still has 90 cents of every dollar he ever earned in his life." I don't know if Daddy was quite that tight with a buck, but he didn't believe in wasting money, that's for sure. No one ever turned on an extra light in our house, and long distance was used for emergencies only.

I can't recall a single time in the 20 years since I left home that my father actually called me. My mother would occasionally call, but only to get something important said, never just to talk.

None of us would do with our own money what is done with the hard-earned tax dollars sent to Washington. I wish I could say I was as careful about spending my own money as Buddy and Helen Watts. That was one lesson I could learn over again. But when I'm charged with spending other people's money, I apply one of the Watts family principles: "Spending money ought to be as difficult as it was to earn it."

When I've opposed increasing the budget for certain programs, some people have criticized me: "Watts is a Republican, but I guess he's forgotten where he came from." They don't understand that where I came from, they'd horsewhip you for spending money the way Washington often does. Where I'm from, if business was as bad delivering services as government or as bureaucratic, they wouldn't be in business long.

By the time I decided to run for the OCC I had begun to understand that while government does need to be restrained, it can still play an active and positive role in society. Of the three commission members, two had already voted to let the phone company keep the excess profits. One of them was my opponent. A Watts victory in the election would tilt the commission toward empowering real people by letting them have some of their money back.

A lot of money was at stake in this race, and when big money is threatened, politics can get downright ugly. I'd taken some grief from my family and friends when I switched parties, but that didn't compare to the criticism I took by becoming a Republican candidate.

I discovered the hard way that just putting an "R" behind your name changed some people's view of you and your ideas, even though I was the same man and my ideas hadn't changed—just my registration. Because of my college football years, I was fairly well known in Oklahoma before I announced for public office. Invitations to speak at sports banquets and school assemblies arrived on a fairly regular basis. Most were great opportunities to encourage kids to be the best they could be, to set their sights high and believe they could achieve their dreams with hard work, sacrifice, and commitment.

When I was conveying that message all across the state, I was a Democrat, and no one raised any objections. To the contrary, my motivational speaking usually got a fairly positive response. Then I became a Republican. Early in my congressional career, the *Norman Transcript,* a staunchly Democrat paper, blasted me in an editorial saying that I *used* to be a hero to the kids in Oklahoma. It went downhill from there.

With a few campaigns under my belt, I now understand that the "politics of personal destruction," as we have come to know it, is a sad fact of life in the political arena, and my skin has toughened up considerably. But at the time, it was difficult for me to understand the shallowness of some criticism, which seemed to be based solely on my choice of political party. If I'd had a big "D" behind my name, I've always wondered if the editorial writers would have been quite so harsh. My first term in Congress, I worked hard to save a defense project in the district—an effort that had been started by the Democrat congressman I replaced. The Clinton administration's Defense Department was trying to block it. We fought to keep it on track and were successful. Yet a Democrat newspaper publisher in my district admitted to me, "If the project had failed, I would have blamed you, although I know it wasn't your fault." He was a true-blue Democrat.

Obviously I didn't fit the mold. Clearly, a black person running for any office in those days was expected to run as a Democrat, and not much has changed. But not everyone was willing to let my new party affiliation change their view of me. During my campaign for the commission seat, I was helped by another friend—a state legislator by the name of Kevin Cox.

Ironically, Kevin is a liberal, black Democrat, but he and I formed a friendship that I value to this day. We disagree on many issues, but not on our ability to respect each other's opinions. As a favor, Kevin took me to a black Masons convention in Boley, Oklahoma, where the organizers were kind enough to allow me a few minutes to speak to the group about my candidacy.

After my remarks, I walked by a lovely black woman, in her late seventies, I'd guess, who was playing the piano for the meeting. "Honey," she said, "I sure did like your talk. But I can't vote for you. I'm a Democrat."

I laughed, having gotten used to hearing that from a lot of black voters I met along the campaign trail. But Kevin was no quitter.

"Wait a minute," he told the little old lady. "You can't vote for him in the primary, but you can vote for him in the general election." Then he smiled at her and added, "Just make sure you go back to the Democrat side of the ballot right after you do." Shades of Daddy and Uncle Wade.

Two or three weeks later, I was talking to a friend from Tulsa about another mutual friend, a young woman in her mid-twenties. She told me that she'd recently asked the other woman if she was going to vote for me for corporation commissioner.

"I'm a Democrat," the young lady replied. "I can't vote for him."

"Not true," my friend told her. "In a general election, you can vote for whoever you want."

The woman was dumbfounded.

"Really?" she asked. "I thought if you were black you had to vote Democrat." And she was as serious as a heart attack.

That first campaign was an eye-opener for me in more ways than one. I was used to seeing politics portrayed in the movies and on television. It always looked like one endless balloon drop surrounded by adoring crowds. I obviously had a lot to learn. Most of the time, I spent what seemed like hundreds of hours alone behind the wheel of a 4-year-old Chevy Caprice busting my way up and down the highways and back roads of the great state of Oklahoma from Guymon to Broken Bow and every other part of the state.

I couldn't afford motels and hotels on these long jaunts, so I drove home every night, often arriving in the wee hours of the morning. Most of the time, it was just me and my thoughts out on the highway, stopping every now and again at a roadhouse or country store for a bite to eat. The loneliness was intense, and I know it was hard on Frankie, with four children in the house and another on the way. She was and is a remarkably strong woman who has managed to keep the house and family on a steady course as I have pursued my dreams.

When the loneliness got the better of me, I'd often find myself wondering, "What's the value of all this? Can one person make a real difference? Will people reject me out of hand because of the color of my skin or my party registration?" My decision to run wasn't an earth-shaking revelation, but in my heart, I knew I was on this journey for a reason, so my instincts led me to believe people would judge me solely on my character and my views on the issues. They didn't disappoint me, and it was an exciting night for Frankie and me when the people of Oklahoma elected me as the first African American to the Oklahoma Corporation Commission and the first to a statewide post. I won my seat on the commission with just over 50 percent of the vote in a three-way race.

Running for Congress may be a more prestigious position, but for most candidates, your first election victory is something you never forget. The excitement of winning, the relief that the long campaign is over, and the gratitude and humility you feel all come together in one rush of emotion that is truly thrilling. I had felt this kind of elation when we had big wins at OU, especially that last-second touchdown in my final Orange Bowl appearance. This was much the same, and I can remember that election night as if it were yesterday.

There are times, however, when I think back to that hard, lonely campaign and wonder how we got through it. One of the most nerve-wracking moments of the campaign was a joint appearance one evening for all candidates running for statewide office. I was the rookie in the group and even with my gift of gab, I was nervous to go up against a group of seasoned political veterans. As luck would have it, I was seated next to an older white gentleman during dinner. We were having a pleasant enough conversation, and though I can't remember for sure, odds are we were probably eating chicken. There were times during that campaign I thought I might actually sprout feathers, given the number of chicken dinners I managed to put away.

As the meal was about to end and the speaking began, the man turned to me and said in a nonchalant voice, "I think what you're doing is great for your people. But even if you win, you're still just going to be a nigger to some folks."

It was a remark that stopped me in my tracks and left me nearly speechless. Giving him the benefit of the doubt, I assumed he meant his comment as a backhanded compliment, so I let it pass. Others might not have been so understanding, and they might well have been justified in giving the man a lesson in racial sensitivity. But I held my tongue.

As I sat there, I couldn't help but think of another banquet I had attended more than 14 years before. With the recruitment season in high gear, I found myself invited to attend a luncheon celebrating the

University of Oklahoma's second national championship in a row. Coach Switzer had asked General Chappie James, the first African American four-star general, to be the keynote speaker. General James's remarks were to have a profound impact on me. I had never seen a black man as imposing as he—well groomed, clean-cut, every hair in place, well over 6 feet tall with broad shoulders that made him seem even taller in his military uniform. To this day, I can't help but get a lump in my throat when I see our military sons and daughters wearing the uniform of their country with pride and dignity. My own daughter Kesha served as a communications specialist in the Air Force and the Air Force Reserve, and Frankie and I could not have been prouder.

General James was a powerful speaker, and I was mesmerized by this man who was the very image of a strong, self-confident leader comfortable in his own shoes. He talked about many things that afternoon, but one story I've never forgotten. He told us that when he was young and heard people call him nigger, he found it difficult to control his anger. As a 17-year-old who had known the same denigrating experience, I certainly could relate to that feeling.

But his father gave him some good advice: "You just stay focused on what you're supposed to be doing. Someday, you're going to accomplish everything you want to accomplish, and those same people are going to be standing on some street corner."

I couldn't help but think again of the McGuire brothers, who used to chase us around the neighborhood taunting us with racial epithets and physical threats. As I got bigger, I got braver—or more stupid, depending on your point of view—and I would yell back at them, just itching for a fight.

It takes a lot to get me mad today, and I don't blow up very often. But I had a much shorter fuse when I was a teenager roaming the streets of Eufaula with my friends. Fortunately, before any of us killed each other, I came to realize that fighting accomplished very little and risked more than the McGuire brothers were worth. General James's speech reinforced my own father's wise counsel to turn the other cheek, and his words have stuck with me over the years.

About the time I began thinking about writing this book, I was on a flight going home from Washington one evening. A spit-and-polish military man, who looked to be in his late forties was sitting in the gate area. He recognized me and walked over to talk for a few minutes before we boarded. He gave me a card, but I just glanced at it and didn't actually read the name. I should have. As we talked, he mentioned his father, and I looked down at his card again. It said, "General James." He was the adjutant general of the Texas National Guard.

"Was Chappie James your father?" I asked.

"Yes, sir," he said. "He was my dad."

I told him what a privilege it had been as a young man to hear his father at that banquet. He grinned and proceeded to recite the speech I'd heard that afternoon over 20 years before—almost word for word as far as I could tell.

"I've heard the speech once or twice myself," he told me, laughing.

The words were just as powerful sitting in the bustle of that busy airport as they had been on that day in Norman, Oklahoma, so many years before. And they served me well at my first big political appearance as I sat patiently listening to my dinner companion put his big foot where his baked chicken should have been. As I drove home that night, I thought a lot about what he'd said, and I never doubted again the wisdom of my decision to seek political office. As far as we'd come in making this a color-blind society, clearly, we had "miles to go before we sleep" which might have been the slogan of my campaign that year in more ways than one.

What I didn't know was that the heat of my first campaign would seem mild in comparison to the fireworks I would face as the newest member of the Oklahoma Corporation Commission.

I heard it said once, "If we could sell our experiences for what they cost us, we'd all be millionaires." If that were true, regarding my time on the commission, I'd be a rich man today. As challenging and rewarding as my tenure on the commission turned out to be, it was also one of the most frustrating and difficult periods for both Frankie and me. I approached my new commission responsibilities with the green enthusiasm of a novice politician.

With my last child, Trey, about to enter the world, my life was about to get even more complicated. Yet despite the new demands on my time, I found my roots with Sunnylane were too deep to simply walk away. So

I remained as youth minister, sandwiching time with the kids in the evenings and on weekends.

Whenever I could find a spare hour in what was a pretty full schedule, I also tried to keep myself in shape. When you eat like I do, you have no choice. Who would think the YMCA would provide the opportunity for another run-in with one of Daddy's bears? But it did. Late one afternoon, I had headed to the Y to do a little weight training and jogging—my usual exercise routine. As I headed around the running track, I watched in disbelief as a man, who probably tipped the scales somewhere near 190 pounds, ran through a group of young kids knocking one fifth-grader flat.

I raced over to the little guy to see if he was okay. I was incensed. I quickened my stride and caught up with the guy.

"Hey, man," I told him. "You could have hurt one of those kids. Don't you know any better than that? That's not how an adult is supposed to act!"

He returned my verbal condemnation with a few choice personal comments of his own. Once again, I held my tongue. But then he did something totally unexpected. He squared off his feet and drew up in a boxer's pose. This guy wanted to fight.

Well, the words of Daddy and Uncle Wade and Chappie James were all forgotten. It may have been a macho bluff, but I had spent three years learning to box. When he leaped into position, something triggered my reflexes. In Golden Gloves competition, whenever an opponent squares off, you know something hard and unpleasant is about to come your way. So without stopping to think, I instinctively shot out a quick left jab. By the time law enforcement arrived, the guy was down and bleeding. Unfortunately for me, the newest member of the OCC, several people were watching it all, including a "helpful" citizen who felt duty-bound to alert the *Daily Oklahoman*, the state's biggest newspaper.

I've done some pretty stupid things in my life, but this was a real doozy. Regardless of how outrageous this man had acted, I was totally wrong to respond in the way I had. I should have marched myself to the YMCA office and reported the man. This was a job for the staff to handle, not me.

Once I had calmed down, reason prevailed, and I knew immediately this was going to be news, and *not* good news. On my way home, I called my chief of staff and asked him to prepare a press release apologizing for my actions. I did the same with the youth group next time we met. The last thing I wanted, as a role model, was to send the message that violence is an appropriate response. I also made a solemn vow never to do anything like that again.

In the midst of all the hoopla surrounding the incident, one reporter asked me if I was going after the swing vote. Funny guy. I really expected people would take me to task for my unauthorized boxing match, but people are very forgiving. Or maybe they simply forgot as I found myself in the headlines again, this time for a far more serious reason.

Early in my term, the commission became embroiled in a nasty scandal that revolved around events that had happened before I was even on the scene. But as I was to learn, when the mud is flying, even the innocent need raincoats. I tried to stand above the rumors that swirled around the commission and its members, but I found myself forced to deal with their impact on my family and my ability to do my job. Even though the actual activities that were under investigation had nothing to do with me, the rumors never stopped, and from time to time, I found myself having to explain and refute the lies over again.

Every reporter who does an Internet search brings up the Oklahoma Corporation Commission scandal, and I have to walk them through the real story. So what is the real story? Early in the campaign, I got a call from an FBI agent who wanted to get together. He told me that as a candidate, I needed to be briefed on potential ethical problems that might arise. He wanted to give me a little free advice on how to be careful about influence peddling and other potential ethical problems. It sounds like a scene from a movie, and sometimes it felt that way, but I agreed to meet him at a local restaurant. After a Coke and what seemed like a standard ethics briefing conversation, we got up to leave.

"There are some people who care about you," he said as we shook hands to say goodbye, "who think you're a good guy, a person of integrity."

In my innocence, I didn't think to ask him who he was talking about or if there was anyone specifically I should be wary of. Warning bells should have gone off in my head, but they didn't. This "briefing" seemed no different to me than the ones I had gotten at Oklahoma University. Coach Switzer had FBI agents come in every year and talk to the team. All the NFL teams and many in the NBA do the same thing to protect their players from making the kinds of mistakes that could lead them or their teammates to disgrace or worse. Even when I played ball in Canada, someone from the Royal Canadian Mounted Police would meet with the team each year, telling us what situations and what kinds of people to beware of.

Drugs and gambling were the issues then, but this was something different—graft—and I now realize the FBI was trying to send me a stronger kind of signal. There were things in the works at the FBI, and before long, they hit the public eye like a sharp chip of gravel. But I was too busy trying to win an election to give much thought to his warning.

Oblivious to the storm forming around me, I was sworn into my first elected position in January 1991. As one of three members of the commission, I would have a chance to influence public policies that touched the economic lives of nearly every person living in Oklahoma. By California or New York standards, 3.5 million people may not seem like such a big deal. But temptations can come with all kinds of power, large and small. As I would soon learn, those temptations had, in fact, gotten the better of some of my commission predecessors, who were being accused of illegally bartering their power for personal gain. Investigators were tracking the situation with all the seriousness it deserved.

One commissioner, a Democrat, eventually served some prison time. Another commissioner, Bob Anthony, a Republican who had been elected in 1988, had been secretly working for months with the FBI to get to the bottom of various influence-peddling incidents that had occurred in 1988 and 1989. With a little more experience under my belt, I would have expected the spillover that was sure to come with a scandal big enough to grab front-page headlines. If I'd known how strong and persistent it would be, I would have dressed for my commission duties in the kind of foul-weather gear people wear on whaling expeditions, because when I joined the commission, I signed up for a shirt-sleeves stroll into the political equivalent of a gale-force storm.

Although the incidents under investigation happened two or three years before I took office, politics is sometimes more about appearances and personal agendas than reality. When I came aboard, just being a commissioner made me the subject of rampant suspicion fueled by a ratings-hun-

gry media. But it was the dynamics at play in a congressional race that actually dragged me into the fray. One unintentional snub turned my life and
my family's life into what seemed to be a nightmare that would never end.

My trouble started when then-President George Bush came to town.
My colleague on the commission, Bob Anthony, was embroiled in a
tough race for Congress, up against a long-term incumbent Democrat.
Not surprisingly, in a state with so many registered Democrats, it was an
uphill battle for Bob, and he was losing ground. When the president
arrived in Oklahoma City, however, he appeared with me instead of
Anthony. I can understand why Bob resented both the visit and me.
Here he was in the fight of his life, desperate for the positive publicity a
presidential appearance would give him. Moreover, Anthony was the
one fighting corruption, and I suppose from his point of view, he
thought he had earned himself a presidential blessing. Maybe he had.

I, on the other hand, wasn't up for election and was the new kid on
the block. I don't know why the president's advisers set up the schedule
the way they did that day. They knew I was a former football star turned
politician and perhaps thought it would help Bush's own campaign to
be seen with me. But whatever the reasons, seeing me smiling from the
front page of the statewide newspapers linked with the president of the
United States didn't sit well with my commission colleague. I had gotten
a lot of attention on numerous accounts—a lot more than I deserved—
and I don't think any of this went over with Bob very well.

After the event, I asked one of my staff to try and smooth his ruffled
feathers by talking to one of his administrative aides. He tried to tell
them I didn't ask for all the presidential attention. *They* asked *me.* And if
there is one thing in politics you learn early on, it is that when the president of the United States wants to meet with you, it's a command performance. Besides, I was flattered and honored to be asked by President
Bush and certainly would have been happy to share the stage with Bob
too, but it wasn't my call.

With the relationship between Bob and me now somewhat strained,
the investigation took a major turn. About a month before the election,
Bob decided to go public at a scheduled commission hearing. About 15
minutes before the hearing was to begin, he walked into my office.

"I've been working with the FBI," he said. "I'm going to make an announcement today. You don't have anything to worry about. None of this stuff relates to you. Just pay close attention to the dates."

This was the first I knew of any investigation, but the words of that FBI agent the year before suddenly rattled back in my brain. When Bob told me his plan, I was shocked to find out this had been going on without my knowledge, but I knew it didn't involve me. I appreciated the heads-up and thanked him for the information. We headed to the hearing that had been arranged to take place inside a courtroom.

What I later found out was that by staging the announcement in a courtroom, under Oklahoma law, Anthony would receive judicial protection for his remarks, as long as he delivered them from the bench. In a dramatic announcement, he detailed serious allegations of bribery and improper communications, providing the gathered media with all his information. He even divulged that he had gone into meetings wearing a wire. The press hopped on the roller coaster and took the public along for the ride. I watched this with amazement, as stunned as anyone in the room that day.

Like any citizen, I was concerned that one of the state's most important regulatory institutions might have been compromised. If illegalities had taken place, if public servants had violated the public trust, then as far as I was concerned, let everything come out and let the judicial process kick in. I was just glad I wasn't going to be part of it.

Little did I know. Three days later, at a press conference, a reporter asked Anthony, "What about the current commissioners?"

He knew that I had nothing to do with the scandal, but instead of clearing my good name, he said, "I can't give anybody a clean bill of health." I was told he went further later on, behind closed doors. "I'm not with the commissioners 24 hours a day," he reportedly said. "I don't know what everyone else is doing." But he knew exactly what *he* was doing, putting the most dramatic spin possible on the allegations to bolster his own congressional candidacy. By creating the illusion that he just might be the only honest guy on the commission, he hoped to give his lagging campaign the momentum he desperately needed.

Anthony's sad attempt at manipulating public opinion created close to two years of pure hell for my family and me and, for the first time in my life, caused me to question the motives of my own government. It's difficult to put into words the feelings you have when you find yourself caught up in a situation not of your own making and without the power to end it. I felt a lot like Richard Jewell. He was the security guard who became the focus of intense media speculation after law enforcement leaks singled him out as the chief suspect in the Atlanta Olympics bombing. Eventually, he was cleared and a new suspect was named, but his life must have been unbearable for a period of time.

Government can make mistakes and so can the media. I was living proof of that in Oklahoma. As the media's insatiable appetite for another screaming headline or titillating story for the evening news continued, so did the stories linking me to the scandal. One particular television station, the NBC affiliate, ran my picture two or three times a week on their 6 and 10 o'clock news, with the headline "Crisis at the Commission." One of the local radio stations was particularly good at keeping my name tied to the scandal even without a shred of evidence. I had gone to journalism school and learned about journalistic ethics, and this kind of journalism was not what I was taught in school. These media organizations took pieces of the story and twisted them into their own version of reality.

One night I saw a segment titled "J. C. Watts's Finances." The financial problems I faced when the oil business slumped were now public information. If my past was impacting the way I was doing my job, then I would have considered it fair game. But they had no proof of any misconduct on my part whatsoever, and yet they used what was a disappointing time in my life to make it appear that I was broke, busted, disgusted, and a prime candidate for corruption.

If every person who has a financial downturn in their life turned to crime, we'd have to double the number of prisons in this country. Frankie and I had gone through a tough financial setback. We were paying our debts, but some in the media used our personal misfortune to create an atmosphere of doubt and suspicion.

Initially, I thought my government would come to my rescue. I knew I had done nothing wrong, so—stupidly as it turned out— I assumed that the FBI and local law enforcement would clear my name. Instead, I found out that the FBI put me under surveillance for about a week. A couple of jokers even had to follow my family and me on a weekend trip to Eufaula. That duty included observing our "getaway" from the driveway of our home in a car filled with kids, crayons, diapers, and car seats. Their surveillance produced a mighty "suspicious" itinerary—our traditional stop at Granny Stephens Fried Chicken, just east of Norman, visits to a couple of shops on Eufaula's Main Street, dropping by Daddy's place, then going to a restaurant and finally to church. I've always pictured those poor agents reporting in on Monday morning, pleading, "Boss, this guy is beyond boring. Please don't make us follow him anymore!"

Paranoia set in but like someone said, "Just because you're paranoid, doesn't mean someone isn't out to get you." I found out that my phone had been tapped, and yet the FBI and the U.S. attorney's office continued to tell me that I was not the focus of their investigation.

Like business, timing is everything in politics. Under Oklahoma election law, as a sitting commissioner you had 120 days after an election to collect campaign contributions to pay off campaign debts. After my race, I still owed money, and not being a rich man, I had to go to potential contributors to help me out. A lobbyist by the name of Bill Anderson had told me to give him a call if I needed help getting rid of the campaign debt. He was an older man, from a rural community, and when we talked it felt as though we had a lot in common. He'd call me from time to time but never tried anything underhanded with me. With the post-election deadline looming and some debt still remaining, I began to do a little fund-raising and called Bill. At one point in our conversation, he brought up an issue that would be before the commission. Delicately but firmly, I told him I would look at the facts and the law and decide. I had no idea the FBI was taping this conversation.

But somehow, Bill's contributions to my campaign became the topic of media speculation. Once again, I tried to be forthcoming; I showed the FBI my contribution reports, and they told me they were satisfied. The media wasn't. The drumbeat got worse.

Seeing the effects the scandal was taking on Frankie and the kids finally pushed me to seek out the local FBI director, Bob Ricks. I told him that Anthony was out of control and so was the press. "J. C.," he said, "I understand what you're saying, but there's not a whole lot we can do."

I then went to the acting U.S. attorney John Green. Behind closed doors, he told me, "We have no interest in you. You have nothing to be concerned about." Nothing to be concerned about? Did he watch the evening news? Moreover, I had become chairman of the commission, and continuous questions about my honesty were beginning to negatively impact my ability to do the job. Bob Anthony continued to drop the FBI's name around the commission staff and public utility officials to intimidate them. I shared this with Ricks. It was enough to create doubt about me even though Anthony knew I was not involved and had done nothing improper.

My anger boiled over. "You see what this TV station is doing to me almost every night. I'm a youth minister. My wife and my kids have to see this stuff. It's wrong to leave me out to dry like this. I want my name cleared."

"We've never had any interest in you," he said. "You can go out and have a press conference and tell the press exactly what we're telling you."

But by now I knew the routine. As soon as the last camera light died, the media would be back on his doorstep, and his office would "neither confirm nor deny" my claims.

During this entire 2-year period, I was never called before the two grand juries that were seated. I was the only person whose name was linked to the scandal that was never put under oath, never deposed, and never asked to testify in any way. Even Bob Anthony eventually said in a deposition, "This isn't about J. C. Watts. I never meant to hurt J. C."

Finally, a former U.S. attorney and friend, Gary Richardson, helped to end what had become my own personal nightmare. Gary told me when he was investigating a public official and the official was cleared, he made it public. "The public's trust in its officials is important," he said. He wrote to the U.S. Attorney demanding a letter stating that there was no investigatory interest in me. Begrudgingly, they finally gave it to me, but the damage had been done. Once tarnished by even the suggestion of corruption, it is hard to completely recover your good name. I am eternally grateful for John Green's decision to provide the letter, which I always suspected didn't win him any points in the office.

After it was all said and done, I wiped the mud off my jacket and kept walking forward. Despite all the distractions, the commission was able to achieve some real wins for the people of Oklahoma. We audited the utility companies and saw rates go down while infrastructures improved. We were able to provide refunds without threatening the good financial shape of the companies. But as proud as I am of this legacy, that time in my life will always be a time of lost innocence and opportunity for me.

Years later, after I had been elected to Congress, I still found myself harboring resentment and animosity toward Bob Anthony for putting me in the path of the steamroller of a scandal he created. They were difficult feelings for me to reconcile with my Christian conscience. I knew I was wrong to continue to have such negative feelings even toward this man who had caused me so much grief. Early in my first congressional term, I was invited to attend a going-away reception for Cody Graves, another member of the commission. I thought a great deal about Bob and came to the conclusion that, despite the harm he had done me, I felt I needed to atone for my own anger for the sake of my soul. So I went to Anthony's office and talked with him.

"I've got to be honest with you," I told him. "When I left the commission, I did not think very highly of you. My heart has not been right concerning you, and that's not the Christian way." Then I asked his forgiveness.

"Oh, no problem, J. C.," he said. "I understand. Hey, we all get there sometimes. I know what you mean." And we shook hands.

My conscience was clear. I had done what I thought I was supposed to do. Later, I learned that at that very time, he was also talking with the state attorney general about me, trying to muster up the same old silliness again.

There are moments with some people that you say, "Lord, how many times am I supposed to forgive him?" In Scripture, it says, "Seventy times seven." So, I guess I've got 489 times more to go.

Media and campaign distortions are epidemic these days. I learned the hard way, don't take everything you read or hear or see on the evening news as the gospel truth. Ratings can sometimes be more important than reality or a person's reputation. I also learned that the government can be a destructive force as well as a positive one. That in its zeal to punish the guilty, it sometimes forgets that the innocent can get caught in the crossfire.

We've seen an increasing number of examples of federal law enforcement abuses that should be worrisome to everyone who believes in justice. Most law enforcement officers are heroes in my book, but our justice system at times can abuse its power at the expense of innocent people. Richard Jewell is a perfect example. I believe I fit the category as well, and having found myself the unwitting victim of our legal bureaucracy, I have to admit that I no longer blindly assume that our justice system is always right or fair. Perhaps a little healthy skepticism about our government isn't such a bad thing for anyone responsible for the public trust.

The viciousness of the political process is a fact that nearly everyone in public office can testify to. Riding the merry-go-round of scandal was no fun whatsoever, but by the time it was over, I was strong enough to handle anything. Next stop: Washington.

11

Independence Can Never Be
Taken, Only Given Away

No man is free who is not master of himself.

Epictetus

In my senior year, Eufaula High made it to the state football playoffs.
Buttons were busting all over Eufaula that fall as the home team
headed up through Muskogee and on to play Locust Grove High, about
40 miles east of Tulsa. We won the game that day and lost to the state
champs the next week in the semi-finals.

When I stepped proudly onto the team bus that afternoon, however,
I didn't know whether we were going to win or lose. And I certainly
didn't know that I would later learn one of the most important lessons
of my life from the events of that day. Given Oklahoma high schools'
reputation for producing first-rank college players, it was no surprise
there were plenty of scouts in the crowd checking out the kids who'd
made a name for themselves on the gridiron that year. It didn't bother
any of us. Scouts had been a fixture at Eufaula games for years, even
before the Selmon brothers.

Just as they had so many times before, my father, Uncle Wade and
several other folks had made the trip from home to cheer on the team. I

could always pick out Daddy and Uncle Wade's wide Watts frames on the sidelines.

This day, a couple of scouts from Texas Tech stood on one side of Daddy and Uncle Wade, shooting the breeze and watching me do my stuff. The McIntosh County district attorney and a few other dyed-in-the-wool Eufaula fans stood on their other side. Like most of the elected officials in the state at that time, the D.A. was a Democrat and proud of it. Buddy and Wade knew him well. They had worn a path to his office door trying to help out folks in their congregations or in the community who found themselves on the wrong side of the law. These were folks who'd made a mistake but, deep down, weren't inherently bad people. The Watts brothers would often go to bat with the D.A. for them.

Everybody knows your business in Eufaula, and justice is meted out with a familiarity that you don't see in big cities. Because of my father and uncle's reputations in the community, the DA tended to listen when they came round to plead on behalf of someone in the community. But this man's cooperation and willingness to show some leniency didn't come free. To put it nicely, he was very patronizing toward blacks. And he was also a little too fond of drink.

That day in Locust Grove, the little hometown cheering section was doing its usual thing—talking themselves up a storm—when one of the Texas Tech scouts remarked on how proud Eufaula must be of its team. The D.A. nodded his agreement, looked across the field, and said, "These are *my* niggers."

Luckily, none of the team heard this remark, but my brother Lawrence had heard it, and I was surprised he didn't deck the guy right on the spot. He was furious.

I didn't find out about the episode until one day several years later, when Lawrence and I were chewing the fat about how politicians take people for granted. He told me the story of the D.A.'s racist comment, and how Daddy and Uncle Wade had remained silent. I couldn't believe it; as soon as I saw Daddy, I raised the issue.

"Why didn't you give him a tongue-lashing?" I asked him.

"One," Daddy calmly replied, "his liquor had gotten the better of him. And two, he's been pretty good about it when we've needed his help from the bench."

I thought for a moment. Usually, I took Daddy's word as gospel. But this time I knew he had given me an excuse, not a justification, and I wasn't satisfied. I was surprised that no one on the sidelines that day, black or white, had given that man the scorn he deserved, and I vowed never to get so dependent on any person or any organization that I wouldn't feel free to set them straight on an obvious injustice. I learned that when it comes down to doing the right thing, the only thing you can depend on is your own conscience.

The great gospel singer Mahalia Jackson said, "It's easy to be independent when you've got money. But to be independent when you haven't got a thing—that's the Lord's test."

My father and uncle may have failed the test that day, but they came of age in a time when nearly every elected official was capable of blatantly racist remarks—or worse. Compared to their Jim Crow, Depression-era experiences, I've had very few hard days in my life. Now, with a little political experience under my belt, I understand far better their decision not to engage this man, but I still can't accept a "go along, get along" approach because I don't believe that giving in on matters of principle gets you anywhere in the long run.

If the black community had stood up to this man just once, if they had told him, "We're going to work against you. You won't take our vote for granted again," that man would have had a life-changing experience; and both he and the black community would have been better off for it.

But in those days Oklahoma was a one-party state. The D.A. probably didn't have a Republican opponent, although there must have been Democrat alternatives. Past favors or not, the black community had no reason to stay hitched to this guy, but they did.

My place, however, wasn't to judge anyone, but rather to learn from the incident, and I did. I had discovered that individual independence is a precious commodity that can never be taken, only given away.

Nearly 15 years later, that experience was to serve me well as I made a decision that would take me even further from my roots, from the small pond of Norman, Oklahoma, to the stormy waters of Washington, D.C.

In the spring of 1994, rumors began circulating that then-U.S. Senator David Boren, an immensely popular Democrat, was thinking about resigning his seat to accept the presidency of the University of Oklahoma. The rumors turned out to be true, and speculation on his possible successor hit the state harder than an Oklahoma twister. It was clear that Republican Jim Inhofe from the First Congressional District and Democrat Dave McCurdy from the Fourth Congressional District would run for the Senate. But then people began approaching me about running as well. When I ran for the OCC post, I hadn't looked any further down the road than Oklahoma City. Now, suddenly, I was faced with two possible opportunities—running for the Senate or for the Fouth Congressional District seat, which was about to open up. The question was, did I want to go to Washington? Did I intend to make a career in politics? Is that where my heart really lay? What about my family? I didn't know the answers.

So I set about finding out. I made a quick trip to Washington and met with Senator Phil Gramm, who was then head of the Republican Senatorial Campaign Committee. He gave me a thorough briefing, but with Inhofe in the race, he took a neutral position on my candidacy. I like to joke that Phil sat on the fence because he was from Texas and just couldn't get past those OU-Texas and Texas A&M games. But, in fact, he had to stay impartial, given his position.

The same couldn't be said for Senator Arlen Specter of Pennsylvania, who strongly urged me to throw my hat into the ring. I also met with Senator Bill Cohen of Maine before returning home to think things over. Frankie and I talked long and hard about the race. In the end, I decided I was ready for Washington, but not for the Senate. I had won a statewide race three years earlier but I didn't feel I could take on another one at that moment with four kids at home. At the same time, it was clear to me that much of the Republican old guard was lining up behind Jim Inhofe's candidacy. I didn't take it personally. Jim had paid his dues and it was his turn.

So I had to make a decision that I knew would still alter my life in a very dramatic way: Should I run for Congress? In 1990, Jack Kemp spoke to the Oklahoma Republican State Convention. I was astonished. He was the first national Republican I ever heard articulate what I was feeling: The Republican Party needed to be a growing, inclusive party. Targeting underserved communities with the economic tools to help people help themselves should be one of our party's core principles.

Welfare was doing more harm than good and needed to be totally reformed. And tax relief was crucial to our country's long-term economic future.

Jack's inspiring words came back to me as I debated whether Washington was the place for me. Could I make a difference? Did I have the right stuff? And most important, could I remain the kind of independent Republican I am and still be effective in the heated atmosphere of partisan Washington?

Frankie and I also talked long and hard about the effect my going to Washington would have on our family. We just weren't sure that this was the right move. I needed to talk to Tom Cole, but I had a problem: My friend, mentor, and political adviser was also considering a run for the seat.

Tom thought that I might be a viable candidate. I knew for sure that he was. Meanwhile, a lot of different people were encouraging us to run and we were both torn. Each of us offered to defer to the other. We decided to decide who would take up the banner over breakfast. I always think better on a full stomach.

As we sat stirring our cups of coffee, waiting for breakfast, a man came up to the table and introduced himself. He remembered me from my days quarterbacking the Sooners and just wanted to say hello. I had barely put a fork into my eggs before another guy stopped at the table, followed shortly by a couple, all just saying "howdy." About the time we got our serious discussion back on track, yet another well-wisher dropped by the table.

"Well," Tom said, smiling and with an odd kind of relief in his voice, "I think we've already had our straw poll. I think we know who the candidate is." And that is how I finally decided to throw my hat in the ring to join the most touted bunch of Republican freshmen in the history of the Congress—and I got breakfast to boot.

I went home and talked once more with Frankie. We still weren't sure, but somehow it seemed as if the hand of the Lord was guiding me down yet another path. The decision was made: I would run for Congress.

The Fourth Congressional District stretches from the suburbs of Oklahoma City to the Oklahoma-Texas border, where the Red River occasionally jumps it banks and plays havoc with state lines. It is an

amalgam of diverse interests that poses problems for a congressman trying to pick committee assignments. The Fourth District is home to a
number of important military bases, including the Fort Sill Army Post
and Altus and Tinker Air Force bases. Tinker is the state's largest single-
site employer, with more than 23,000 military and civilian personnel.
The district is also heavily dependent on agriculture and energy. The
University of Oklahoma, located in Norman, is also in the district. This
unusual economic mix has made defense, education, agriculture, and
energy all key local issues in this wide-ranging district.

A moderate to conservative area, it is overwhelmingly white (84 percent), with small African American (9 percent), Hispanic (4 percent),
and Native American (3 percent) populations. People there had sent
Democrat Dave McCurdy to Washington for 14 years, but George Bush
won the district with 58 percent in 1988 and 41 percent in 1992. With
the more liberal areas surrounding the university, moderate to conservative Democrats in the rural areas, and a growing Republican suburban
influence, there was no question that the Fourth District was a swing
seat.

I knew it wasn't going to be a cakewalk, but I didn't go into politics
intending to lose. That wasn't overconfidence, just the plain old-fashioned
determination that I learned from my daddy and my coaches. I never went
into a football game I expected to lose. And I can say the same thing about
every one of my election campaigns.

First, however, I had to get the nomination, and that meant beating
out four other candidates in a tough August primary. I managed to get
49 percent of the vote—1 percent short of winning the nomination outright—so state representative Ed Apple and I squared off for one more
round. A businessman with a military background, Ed was a good candidate and a good guy. It was a fair fight, but a month later, I came out
on top by four points—a little closer than I like 'em.

With my primary behind me, my focus now turned to my Democrat
opponent, David Perryman, a lawyer and former county chairman. The
campaign settled down around several issues—defense, term limits, gun
control, abortion, taxes, and, of course, the Contract with America.
While my race for the Oklahoma Corporation Commission hadn't been
easy, it was in this campaign that I first experienced the willingness of

the opposition to embrace the kind of "anything to win" tactics that characterize our elections more and more today. During the primary, the National Rifle Association asked all candidates to fill out their election issues questionnaire. Perryman's responses earned him an F from the organization. My answers received an A.

I went into the general election feeling pretty confident of my position with the local NRA and its members. What I hadn't counted on was the political pressure that would be put on the national NRA in Washington by Perryman's supporters.

By waffling and backtracking a little but without really changing his position on any issues, my opponent miraculously earned himself not just a passing grade but a full-fledged A on the NRA's general election report card. I remain a strong supporter of the Second Amendment, but believe me, the NRA has little to do with it. I was terribly disappointed in the NRA.

The Contract with America, which covered everything from welfare reform to a balanced budget, became a major issue in the campaign. When asked by a reporter about the Contract, Perryman said he thought it was a pretty good idea and even went so far as to say he wished his own party had done something similar. That was before he went to Washington for campaign meetings with Democrat Party officials. He came back from his few days of "political education" staunchly opposed to the Contract—another one of those political conversions on the road to Damascus.

Despite his flip-flopping on issues, the race remained neck-and-neck. I suppose I should have expected what came next, but I have to say that I was as surprised as the next guy to see my opponent decide to play the race card. After a statewide election and 4 years on the commission, I wrongly assumed that the color of my skin was no longer an issue. More than that, I was brought up in a family where we learned early on that the Democrat Party was on the side of black Americans. That led me to believe that the last person who would use race to win an election would be my Democrat opponent.

I was never more wrong in my life. Perryman began running TV ads with a picture of me from the days when I wore my hair in a full-blown Afro, the style my father always hated.

For many people, the hairstyle still conjured up visions of Black

Panthers and black power, radicals and revolution. If they had only known that in my younger days, my Afro had been more about being cool than confrontational. It didn't take a political scientist to see my opponent's use of this outdated image was a not-too-subtle appeal to racial fears. The underlying message was clear: "Do you want some bushy-headed troublemaker representing you in Washington?"

Back then, I was also surprised when the NAACP, for which my uncle Wade and my father had worked so hard for so many years, didn't raise its voice to object to the ad. (I wouldn't be surprised today.) While the Democrat Party defended it, the national press was silent, with the exception of the *Wall Street Journal* and, believe it or not, the columnist Bob Novak. They both condemned the ad. Another lesson learned: When it comes to the NAACP, being black and the target of racist politics guarantees you nothing unless you adhere to their narrow ideology. For them, the retention of political power supercedes skin color.

But I did have my defenders. Kevin Cox, the liberal Democrat state representative, who had helped me win the Corporation Commission spot, launched an effort across party lines to denounce Perryman's tactics. He and a group of black ministers held a news conference to condemn the ad as a racist attack and nothing more. I have never forgotten their courage and honesty at that crucial moment in the campaign. The Democrats' decision to use racism backfired when people rejected this kind of politics.

Even though only 9 percent of the Fourth District is black, and 68 percent Democrat, on election day, my opponent limped in 9 percentage points behind me in a three-way race.

There was more stress to come in my political life, but when the flag of racism was flown, the innate goodness in people came forth. That gives me hope for our country and the direction it's headed on racial issues. More than that, remembering that turning point in my first congressional campaign always sustains my spirit when the going gets tough in Washington.

It proved to me there are a lot of good people in this country who just don't want to be associated with negative politics. There are plenty of white people ready and willing, even eager, to vote for a black candidate who shares their values, regardless of political party.

After a grueling campaign, on election night 1994, I won with 52 percent of the vote, becoming in the process the first black Republican elected to Congress from a southern state since Reconstruction. But it was a historic night in more ways than one. I was also part of the first Republican majority in the House of Representatives in 40 years, and that was a thrill for all of us.

After an election night victory party filled with plenty of handshakes and hugs, Frankie and I headed home to put our feet up and just look at each other. Suddenly, the last scene of the movie *The Candidate* popped into my head, the moment when Robert Redford, who has just won a Senate seat, panics and asks his campaign manager, "What do we do now?" I think that's a little how Frankie and I felt election night. Now what?

We had already discussed what the family would do if I won. Frankie and the kids would stay in the district, where friends and family were close.

I would commute between Washington and the district, knowing I would be back home every weekend doing congressional duties anyway. In retrospect, that decision was probably a mistake. While I have gone home every weekend to attend events and be with my family, I underestimated how much time I would miss with my kids and Frankie during the week. A lot of members of Congress attend receptions and dinners on weeknights after the day's legislative duties are over. That's not my cup of tea. I'm a stay-at-home kind of guy, so with no family around, I usually found myself either at the office working late with my staff, doing a little exercise to stay in shape, or hanging around my apartment catching up on my reading.

The first couple of years, I think most freshman congressmen and women run on pure adrenaline. There is so much to learn, so much to do, shoring up political support and spending time with constituents. I was so busy I didn't have time to think about what I was missing. Having Frankie and the kids stay back in Oklahoma may have made sense the first term, but I wish we'd reconsidered our earlier decision after my reelection.

But back in 1994, we didn't know what we know now. Then, all we knew was that our lives were about to be turned upside down, and we wanted to keep our children's home situation as stable as possible amid the bedlam that was about to hit. The first big event was my swearing-in as a member of the 104th Congress.

Big plans were under way for the family to head to D.C. for the ceremony, but this wasn't going to be the usual swearing-in. After it became an issue in the campaign, I had promised not to resign my Corporation Commission seat until a new governor was sworn in. The outgoing governor was under a legal cloud at the time, and many people felt the incoming governor should appoint a new commissioner to fill my seat. This wasn't a problem for me, but it did pose a problem for Newt Gingrich because it meant I could not be sworn in with the rest of the Congress on January 5. The soon-to-be Speaker understood my dilemma and offered to do a special swearing-in ceremony for me on the House floor four days later, on January 9.

We all headed to Washington for the big day—Frankie and the kids, Daddy, several other relatives, and friends. Unfortunately, Uncle Wade couldn't make the trip. The morning of January 9, I had some paperwork to take care of before the ceremony, so I went on ahead of the gang to the Capitol. Frankie would follow shortly with all the kids in tow. On paper, this plan made perfect sense. But we were definitely country mice in the big city. We were used to Norman traffic, not the craziness of Washington, D.C., and sure enough poor Frankie got caught in one of Washington's famous traffic jams. Meanwhile, I stood pacing in Newt's office watching the clock tick toward 2 P.M., when I was scheduled to take the oath of office in the well of the House with my colleagues watching. We waited until the last possible minute, but finally, Newt told me that we had to go ahead.

As Newt and I stood together in that grand chamber where so many historic events have taken place, I felt a lump in my throat. Like most Americans, I'd only seen it on television, and I have to admit to being just a little bit overwhelmed. But I managed to get through the oath to become the newest member of the new Republican majority Congress.

I had a few butterflies in my stomach that day, not knowing what to expect. I really had no clue. I thought being a member of Congress would be interesting, challenging, and, hopefully, rewarding. I had no idea that the constant pace and pressure would come to remind me of the days when I was running for the end zone with a pack of linemen hot on my tail.

When Frankie did arrive, Newt graciously offered to do another swearing-in ceremony in his office, and this time the family was there to see it. I felt extremely grateful to have Frankie and the kids there with me, considering that they had sacrificed so much to help get me there, and I still feel this way today. Daddy didn't say too much, but I know he was proud. I only wish Mama had been there to see it, too. I always laugh when I think about Daddy's one and only visit to Washington. He loved to tell people how much he hated our nation's capital, with its traffic and crime, but most of all he liked to talk about the pancakes.

You see, the morning of my swearing-in, Buddy wandered down to the hotel restaurant for his usual big breakfast. He ordered a stack of pancakes and ate them on down. But when the bill came, I thought Buddy's old heart was going to give out right there in the dining room of the Dupont Hotel.

"Nine dollars!" he cried. "Nine dollars—do you know how many pancakes I could whip up back home with nine dollars!"

I don't think he ever got completely over that breakfast. He talked about the price of pancakes in Washington for years.

For me, the day was a whirlwind of new faces and old friends. Everywhere I turned, there was a reporter dogging my steps, and I was still having trouble just finding my way around. When Frankie and the kids headed home, I found myself coming face to face with a whole new life and whole new set of challenges. "Am I ready?" I asked myself.

My first day on the job felt a little like my first football practice at OU. Deep down, I believed I had the necessary skills. Now, I had to ensure that I also had the mental focus and the right attitude to get the job done. I had always been an independent cuss from the time I was small. Sometimes, that independence got me into trouble, and it still does—I'll tell you how in the next chapter—but as I began my congressional career on a cold day in January 1995, I thought of another football field and what a lack of independence cost my daddy and my uncle. I swore to myself that whatever came my way in this wild thicket of partisan politics I had just entered, my independence must remain inviolable.

Teddy Roosevelt, whose very life was a symbol of independence, said it this way: "A man of sound political instincts can no more subscribe to the doctrine of absolute independence of party on the one hand than to that of unquestioning party allegiance on the other." In the years to come, I was to find the truth of those words.

12

Independent Spirits and Political Ideologues Are Seldom Compatible

They will say you are on the wrong road if it is
your own.

Antonio Porchia

When I arrived in Washington, I was thrilled and excited, but I probably confused more people than anything else those first few weeks. Folks from Capitol Hill to K Street, lobbyists' territory, were trying to figure out whether I was a round peg or a square one so they could neatly put me into the proper hole. That characterization process happens to all freshmen, but a black conservative? That's a little harder to figure out. Office of Personnel Management director Kay Cole James says a conservative is "someone who believes, and who acts like they believe, what their grandma taught 'em." Mittie would have liked Kay.

The truth is, I just didn't seem to fit any of the traditional congressional types. First, I didn't join the Congressional Black Caucus. Reporters were especially confused by that. If you were black, of course you joined the Black Caucus. It went without saying, but I went the other way.

I saw the caucus as a vehicle for group identity, something I had rejected my whole life. It wasn't that I didn't share many of the same concerns as the caucus members or that I was somehow rejecting my roots. Many of our goals were the same, but I feared the increasing demands put on blacks to toe a particular ideological line would manifest itself tenfold in this particular caucus. I wanted to remain a free agent. I had my own views and strong principles that I simply could not and would not compromise for group unity—any group, as my fellow Republicans have also found on occasion. Moreover, I felt that the Black Caucus had become an extension of the Democrat Caucus and too often had been used as a tool for partisan political gain. I may have been only one of two black Republicans in the House, but I wasn't going to trade either my principles or party for membership in the group identity.

In years to come, I believe this country will see many more black conservatives in the public arena as politics in the black community mature and broaden, much as they have in other communities. Ethnic politics are as American as apple pie. There was a time when we thought in terms of *Irish American* politicians or *Italian American* politicians. Nowadays, they're just *politicians*—who are also Irish American or Italian American. It is a distinction with a difference. Their ethnicity is still an important part of their identity, but it is no longer all that they are.

I hope my election has signaled the first wave of politicians who are African Americans, not African American politicians. It is a natural evolution that we are seeing in both parties, though not as fast as I would like. The late Ron Brown, the former secretary of commerce in the Clinton administration, was a Democrat who was black, not a black Democrat. He didn't allow himself to be pigeon-holed in a "minority" cabinet spot. He wanted Commerce because he believed in himself and his ability to lead American industry. Ron Brown refused to fit neatly in one category. He valued his independence and was as successful a politician as this town has seen. Supreme Court Justice Clarence Thomas and Secretary of State Colin Powell are two more good examples of individuals whose blackness is an important part of their identity and to each of them personally but doesn't define them. The experience, values, integrity, and talent of these two men are just as important as their ethnic heritage, if not more so.

Because I was a black Republican, some people in Washington probably viewed my arrival with some cynicism and saw me as nothing more than a token. There was nothing I could do about that except prove them wrong. Others were uncomfortable with the fact that my faith is such an integral part of who I am and what I believe. I couldn't do anything about those folks either except try to make them see over time that faith and public service are not mutually exclusive.

The fact that Washington is often one step removed from reality doesn't make the effort any easier. Members of Congress are under constant pressure to shape their views to meet the expectations of a seemingly endless parade of interests. That includes the interests of our constituents and the many groups that represent them in Washington and, in a broader sense, the interests of our nation. It's often a high-wire balancing act as these competing interests, some within our own districts, vie for our support.

But Washington also offers a challenging and stimulating opportunity to do good, and I realized that from the very moment I arrived. I had been appointed to the Armed Services and Banking Committees. With many military bases on the endangered species list, Armed Services was a perfect fit for me. My district is heavily dependent on its military bases and the jobs that go with them.

Banking turned out to be a real baptism by fire as we took on the Whitewater issue. As one of the low men on the totem pole, however, I listened and learned more than I contributed those first few months—one of the smarter things I did my freshman year. I also made one of my first friends in Congress in the Banking Committee, a terrific congresswoman from New York, Sue Kelly. As anyone who has sat through a Banking Committee hearing will tell you, the testimony can get a little dry sometimes.

As freshmen Republicans, Sue and I sat next to each other at the tail end of the seating assignments. I soon discovered that there was a direct correlation between the dullness of the speakers and the depth of my appetite. The more boring the testimony, the hungrier I'd get. It wasn't long before I'd lean over to Sue and complain, "All this talking is making me hungry."

As luck would have it, Sue Kelly carried a purse that was something akin to a 7-Eleven. She always had something to munch on, and she'd open her bag and sneak me a snack. It's a good thing I didn't stay on that committee: I gained a few pounds that first year sitting next to Sue.

For all of us, first-termers and long-termers alike, the first 100 days of the 104th Congress will go down in history as the most driven, fast-paced opening to a Congress anyone had ever seen and probably ever will. During the campaign, Republicans had promised the American people we would take up all ten items in the Contract with America in the first 100 days of the session. We kept our promise. We combined forces to pass eight of the ten "planks" in the Contract. Eighty percent of the Contract with America is law today. We also term-limited committee chairmen, reduced the number of committees, forced Congress to live by the same rules and regulations as the rest of the country, and cleaned up the House finances.

For a freshman who didn't know anything else, the pace was extremely tough. I was grateful Frankie wasn't there. As patient as she can be, she would have been as frustrated as most of us were with the lack of family time. But a promise is a promise—Newt was determined to get the job done, and he wasn't going to let *us* let the nation down. We worked all hours of the night and day. A couple of times, we had sessions that lasted around the clock. We would recess at about 8 o'clock in the morning, go home and shower, and be back two hours later, ready to go. We wore everybody out, including our staff, the media, and ourselves. Majority Leader Dick Armey joked afterward, "If we ever do a Contract with America again, we should make it clear that we mean 100 days, not 100 nights, too."

We didn't get everything done we'd wanted, but we felt as though we had the world by the tail. I was as proud to be a part of that historic Republican majority as any team I had ever played on. After seeing Democrat Congresses move at a snail's pace for 40 years, accomplishing little more than bigger and bigger government, the first 100 days taught everyone a lesson. Going fast may mean taking a few falls, but it's better than going nowhere. And we were able to achieve what we did for three reasons: we had the American people behind us; we had the will to get the job done; and we acted as one cohesive team putting personal ambitions and antagonisms aside—something I wish we did more of now.

But the exhilarating days of January gave way to the grimmest of realities in April. There are days in our memory that are so extraordinary that we never forget exactly what we were doing at that moment: the assassinations of John Kennedy and Martin Luther King, Jr., the day Neil Armstrong walked on the moon, and recently the day of the World Trade Center and Pentagon terrorist attacks.

On the morning of April 19, 1995, I had attended the mayor of Oklahoma City's annual prayer breakfast. From time to time, when I was in downtown Oklahoma City, I would drop in on a good friend, Clarence Wilson, the general counsel for the regional HUD office. He worked in the Murrah Federal Building.

Clarence and I had been friends for more than 10 years after I'd gone to him for information on HUD loan policies. He was a kind and patient man, always willing to help, and I'm sure that morning was no different. If I know Clarence, he was probably at his desk at 9:02 getting ready for another busy day. He never saw 9:03.

If I hadn't been scheduled to testify that afternoon at a congressional hearing in Dallas on proposed military base closings, if I hadn't caught the 8:45 flight to Texas, I might have been sitting there with Clarence having a cup of coffee.

But I wasn't. I was on a plane heading south, contemplating my testimony on behalf of the military facilities in my district, when a gentleman came up to my seat. He asked if I was Congressman Watts. When I nodded, he said, "I just talked to my wife on the phone. She said that about 9 o'clock there was a gas main or something that exploded in downtown Oklahoma City. She said it sounds pretty serious."

I was as concerned as anyone would be about a potential accident, but we were all to learn this was no accident. When we landed at Love Field in Dallas, usually a busy airport, it almost seemed as if time had stopped. Instead of the hustle and bustle I expected, people were just standing around TV monitors paralyzed by the terrible pictures on the screens. Then, I realized it was Oklahoma City we were seeing. My first reaction was horror and disbelief. Could this really be happening? I had just left there an hour before. Everything was fine. Now, I was slowly beginning to realize that nothing would ever be the same again.

Terrorism had struck the heartland for the first time. It wasn't a gas main, but a terrorist who had taken away 168 lives, and our innocence in the process. I was torn between going back home immediately or staying to testify. Saving our local military bases was crucial to my state, but everything in me wanted to catch the next plane back.

We managed to reach the committee chairman, and he agreed to let us testify early, so I could get back as quickly as possible. When I arrived at the site of the Murrah Federal Building, rescue operations were still going on. The scene is difficult to describe: Firefighters and police were rushing to rescue anyone still lost in the rubble of broken glass and twisted iron, frustrated that they couldn't do more. Other people were just standing and staring at the almost total destruction of a nine-story building. Weeping family members and friends didn't know what to do or where to go. I prayed that in my lifetime I would never again see what I saw that terrible spring afternoon. The destruction was beyond belief. I almost pinched myself a couple of times to make sure I was in Oklahoma City and not Beirut.

It was brutal. Nineteen children killed—many just toddlers, precious lives Timothy McVeigh later coldly called "collateral damage." It was just by the grace of God that anyone survived. When I got there, rescue workers had already taken 167 bodies out of the wreckage. Later they would find one more person who was still alive, and one more body.

I remember just standing there staring at the massive crater where people had once worked and laughed and helped so many of their fellow citizens. I thought of Clarence and other fine people I knew in the federal offices. Suddenly, it sank in. We were just 20 miles from my home, where I had always thought Frankie and the kids were perfectly safe. Terrorism didn't happen in Norman or Oklahoma City. That morning I might have been in the Murrah building, but once again, the Lord had kept me safe. I was shaking as I uttered a prayer of thanks.

A lot of prayers were said in the hours and days that followed the explosion as rescue operations went on; and although the bombing was the product of the worst in a human being, people around the world saw the best as well—the incredible heroics of the firefighters and the police and passersby who risked their lives to help save the injured. We saw the great American spirit of neighbor helping neighbor take hold as the Salvation Army, the Red Cross, and hundreds of volunteers rushed to help the victims and their families.

The fascinating, hope-inspiring thing was that in the middle of all this carnage, we could also see what happens when we put aside colors and labels. There was no red or white or black or yellow or brown, no liberal or conservative, Republican or Democrat. There were just people in trouble who needed help. It was just one America. And the love that poured into Oklahoma over the next few months, from all over the country and the world, was just overwhelming.

There's a poem by Theodore Roethke that begins, "In a dark time, the eye begins to see." In this very dark hour, I saw this country at its best. Moments of great victory or terrible tragedy make us pause and reflect on our place in God's plan. Innocent death can make us question him and ourselves. Twenty-four hours before, I was riding high as a brand-new congressman involved in what was a historic change in our government. Now I found myself angry and frustrated as I reflected on how powerless we can be in the face of pure evil. Like everybody else, I tried to help where I could. My staff kicked into action to provide whatever assistance we could. I quickly found that my minister's hat was a better fit that week.

I had learned at Sunnylane that whenever there is a death in a family, it is difficult to find words that bring any true comfort. But the loss these people had experienced was incomprehensible. This kind of slaughter didn't happen in this country, especially to our children. I can't imagine losing one of my children to a terrorist bomb. The strength of these parents in their grief gave us all faith that we would one day recover.

I went to Clarence Wilson's funeral and to another for the mother of a friend. When I returned to Washington, I realized the exuberance I had felt just a few days earlier as a new Congressman had turned bittersweet. I would always remember the thrill of moving our legislation, the excitement of the first 100 days. I had been a part of history in the making, history that will be remembered positively in the years ahead. But all the celebration and spirit had disappeared into the smoke and dust of that terrible day in Oklahoma City. And for a long time, I was haunted by the horror of the bombing and what it said about the darkness that can grow in the human heart. How does one explain a Timothy McVeigh? Was he evil or just insane? How could anyone take all those innocent lives?

Now, he's part of history, too, and he took leave of us without ever expressing regret or remorse, not even for the children. For me, he will always remain an evil enigma.

It would be several years before Oklahoma City returned to normal; some people's lives will never be the same again. In Washington, we went back to work, soon slipping back into our partisan ways.

Samuel Johnson said, "Sorrow is the mere rust of the soul. Activity will cleanse and brighten it." After the funerals, I threw myself back into my congressional duties and the routine that was to become my life for the next 7 years: Four to five days in Washington each week, weekends home in the district. But somehow, in the aftermath of that dark day, we never recaptured the momentum begun in those first 100 days. Congressional Republicans and the White House clashed over budget priorities and the best way to spend the money the American people entrusted to us. As the months wore on, the bickering got worse. Republicans sent President Clinton several budgets. He refused to sign them. Instead, he shut down the government without regard for the federal employees whose lives were disrupted or for the American people who depended on the critical services government provides. When it came to the budget, it was his way or no way. Compromise was never President Clinton's strong suit, and in the aftermath of the Oklahoma City bombing it was easy to tag Republicans, traditionally the party of smaller government, as anti-government extremists.

The Clinton spin machine went into high gear while Republicans, lacking a cohesive communications message, failed to get our side of the story across. Democrats painted Republicans as heartless extremists happy to weaken Medicare and Social Security for seniors, cut school lunches and education for children, and sell out the environment. President Clinton was always persuasive, if nothing else, and with the media all but collaborating in the White House public relations offensive, Republicans found themselves blamed for what was clearly a Clinton decision to shut down government. In the end, we reached an agreement, but Republicans paid a high price for principle that year and learned a hard lesson we wouldn't soon forget.

As in football, after my rookie year in Congress, my self-confidence began to develop as I became more comfortable with the system and got to know my colleagues better on both sides of the aisle. The summer of 1996 proved to be one of the most pivotal for me, both as a politician and as an African American.

Within weeks of each other, I was asked to make the most important speech of my life and cast one of the most crucial votes of my career. Congress would at last take up the issue of welfare reform, and I would take the platform in San Diego to speak to the Republican National Convention.

The passage of the Republicans' historic welfare reform legislation in 1996 was one of the proudest moments of my congressional career. Republicans had been talking about reforming the failed welfare system for decades, but one Democrat-controlled Congress after another refused to even acknowledge what had become obvious: the system was doing more harm than good to poor families and had been doing so for a long time.

In 1970, Ronald Reagan went to a national governors' convention and proposed welfare reform. He was voted down 49 to 1. Twenty-seven years later, a Republican-controlled Congress dragged a Democrat president kicking and screaming to the signing table. Over the objections of the Democrats in Congress and many of the nation's national black leaders, welfare reform was signed into law in August 1996 to help restore dignity and dreams to those whom prosperity had left behind.

This was a system that was enslaving our people a second time, but many black leaders saw my vote to abolish it as another betrayal in the long battle for civil rights. I saw it as a matter of right and wrong. It was wrong to encourage nonproductivity, wrong to encourage sleeping until noon every day and getting paid for it. It was wrong to tell young girls they would pay no price—moral or economic—for giving birth out of wedlock. It was wrong to penalize poor moms for saving money and poor people for owning homes. But that's exactly what the welfare system, implemented as part of the Great Society programs of the 1960s, was doing.

My Democrat friends didn't want to admit that welfare had failed miserably. They had too much political capital invested in the system to be bothered by facts like the 270 percent increase in the number of children receiving welfare benefits between the 1960s and the 1990s; or the 450 percent increase in the number of out-of-wedlock births. The failure of welfare had taught us that a monthly check is no substitute for making something of yourself, and work not only brings income to a family but dignity and self-respect as well.

At the time, however, many characterized welfare reform in cold, uncaring, heartless terms. Supporters were accused of having no compassion for those in need, especially children. Yet, for most of us, it was precisely the children trapped in an endless cycle of poor, single-parent homes and poverty that drove us to reject the status quo. And today, by any measurement, welfare reform is working. In 1995, 4.9 million people were on welfare; today that number has been cut by nearly half, to 2.6 million.

Contrary to what welfare supporters would have us believe, the best way of measuring compassion is not by how many people are on welfare or AFDC or living in public housing. True compassion should be defined by how *few* people are on welfare because we created a path to success for those willing to work for it. Today, millions of people who were once on welfare are working or in a work program, a training program, or school.

The commitment to overcoming poverty in the old welfare model was measured by the size of the line item in the federal budget, but compassion can't be counted in dollars and cents. It does come with a price tag, but it isn't the amount of money we spend but the depth of the love we give. Compassion means being able to see people as they *can* be, not as they are. Scripture tells us that the measure of a man isn't how great his faith is, but how great his love is. I'm not suggesting that love alone will put food on the table or pay the bill for the kids' school clothes. Nor am I saying that government can't be a positive force for change. But the willingness of one human being to help another find the good in themselves, find the job that has eluded them, find the moral strength to reject a lifestyle of drugs and casual sex—that's what real compassion is all about.

Relying solely on government to address social issues often does little more than exacerbate the very problems that many multimillion-dollar federal programs are intended to solve. It was this message of hope and compassion I took to San Diego two weeks after the welfare reform bill passed. I was privileged to make my "maiden" national political speech before a packed house of Republicans gathered in the Golden State to nominate Bob Dole for president.

I have to confess I was just a little nervous but tremendously honored to have been asked to speak. No one told me what to say or how to say it, and I appreciated that, too. I told the delegates, "In my wildest imagination, I never thought the fifth of six children born to Helen and Buddy Watts in a poor black neighborhood in the rural community of Eufaula, Oklahoma, would someday be called congressman. But then this is America . . . where dreams come true."

I talked about some of the committed souls I had met whose lives of compassion should inspire us all to do more. And then I leveled two challenges.

First, I asked the young people in the audience to fight for what is right in America, to be leaders, to find the courage to say no to things that make them weak and yes to those that make them strong. Then, I challenged parents and adults to help our children be the best they can be; to nourish and encourage character; to boldly stand for what is right and against what is wrong. It was a wonderful evening for Frankie and me, and I remain grateful to Bob Dole for the opportunity he gave me to say my piece.

My strongly held belief in the power of compassion and commitment to make a difference in people's lives had led me into politics in the first place and, as a first-term congressman, spurred me to introduce the American Community Renewal Act. This package of bills was designed to revitalize poor areas so that all people could share in the prosperity that has buoyed the rest of the country. Congressman Jim Talent, the head of the Small Business Committee, and I cosponsored the legislation that was supported by people in both parties (New York Congressman Floyd Flake, a Democrat, was another cosponsor), and elements of the bill could be found years later in the "vision statements" of both presidential candidates in the 2000 election.

For me, community renewal became a matter of conviction and commitment for the next several years, but passing the legislation turned out to be far more difficult than I had ever anticipated in my freshman term. In my naiveté, I assumed that the positive reaction our bill received initially from both sides of the aisle was all we needed for quick passage. Instead, it became a long process bogged down by partisan infighting, the natural snail's pace of most legislation, and the impact of special interests as the bill was developed, amended, and amended some more.

With a Democrat in the White House and Republicans controlling Congress, our legislation met the fate of a lot of good ideas—delays at every step of the way. When it finally became law in December 2000, our accomplishment was more thrilling than any touchdown pass I'd ever thrown, even in the Orange Bowl. Here was something that would make a difference in thousands, perhaps millions, of lives by giving communities and the people living in them an honest chance to succeed. That's what I had come to Congress to help do.

Later that fall of 1996, I won reelection to the House with 58 percent of the vote. I was glad to have those first two years under my belt, and as I was sworn in for my second term, I looked forward to more progress with a growing sense of self-confidence and independence that was to be tested in the coming two years.

My first challenge came just weeks after the 105th Congress was gaveled into session. Apparently, my speech to the Republican Convention had earned me the notice of some of the Republican leadership. But no one was more surprised than I when Newt Gingrich called me one morning to ask if I would be willing to deliver the Republican response to President Clinton's State of the Union Address. Willing? Was he kidding?

For me, speeches have always represented both challenge and opportunity. I've been up in front of enough groups, going all the way back to sports banquets and church congregations, that I don't usually get a case of rookie nerves anymore. But I do wonder if I'm making the most of the moment. A nationally televised address speaking in response to the President of the United States, however, was a different kettle of fish altogether. The stakes were considerably higher. I was being given the opportunity to talk to millions of people with the potential to influence their thinking as well as touch their hearts. I was being asked to speak for my party, defend its vision, and convince the American people that Republicans offered a better way. That's a pretty heady assignment for a midshipman just getting his sea legs and with just three weeks to write the speech of a lifetime.

My staff and I huddled on the best way to approach the actual writing of the speech. Contrary to rumors at the time, the only thing Newt asked me to include in the remarks was a plea for the balanced budget amendment that Republicans were pushing hard. Being a big supporter, I had no problems with his request.

With such a short time period in which to work, we brought on board a solid professional, one who had written beautifully for a number of prominent Republicans. The first draft arrived on the afternoon of February 3, and it was a fine piece of work, but it just wasn't me. Most Republicans would probably have been comfortable with it, but it wasn't my voice. Back to the drawing board.

I had 18 minutes to open my heart to the American people, and I wanted to do it absolutely right. And by now, I only had 18 hours to get it ready. I can imagine some of the more seasoned political veterans' reaction when they heard I had rejected the first draft and intended to write it myself. It may not have been panic in the streets, but that streak of independence I had managed to keep in check my freshman term, emerged at just the wrong time from their point of view. Being the professionals they were, however, they took it in stride while I got down to work.

Because of my many years of public speaking, I had confidence in my ability to put together the kind of speech that would have impact, but it never hurts to get a little help from the Almighty. I called some of my closest friends around the country—both black and white—many of whom were pastors. I picked their brains for ideas and then asked them all, "Please say a prayer or two for me. Ask God to give me the right words."

While the speech was meant to be a partisan response, I didn't want to be seen as a Republican or a Democrat but as an American sharing my thoughts with the American people, who I hoped would understand my heart as well as hear my words. This was a political opportunity, but I saw it as a spiritual one, too. One after another, my friends offered to pray that God would inspire me. I was incredibly moved by their concern; and while some may find it hard to understand, I believed then and do today that I could feel the effect of their prayers during that long night and next day.

However, outside my circle of friends and ministers, there were some curve balls coming my way, too. In crafting my speech and preparing for its delivery, we tried to anticipate any and all circumstances that might arise that night, from a "presidential" surprise to the teleprompter going down. What I didn't plan on was O. J. Simpson. Here was the ultimate irony. On the biggest night of my political life, I found myself in direct competition with another black man, also known by his initials, whose first fame also came from a football career.

As luck would have it, that night O. J. Simpson's civil case had just concluded. His acquittal in the criminal trial had occurred several months before. But that night the public announcement of the civil verdict was imminent. There was every possibility that O. J. would upstage both the president and me. Meanwhile, there I was primed and ready to go. The 60-minute countdown was on. When Bill Clinton delivered a speech, we'd all learned this: President Clinton liked to talk, and when he talked, time really was money. The longer he went, the bigger the budget got. Meanwhile, as the president expounded, I waited nervously in a holding room at the Library of Congress with a few friends and staff members.

I didn't mind the wait. I just hoped he would finish before people had to go to work in the morning. But the networks were in a real bind. Should they cut away from the president or me if the verdict came during either of our speeches? If they didn't, would they lose out in the ratings game to their competition?

One of my staffers asked a CBS executive nearby, "How do you plan to handle this?"

"We don't know," he admitted. "Maybe a side-by-side or a break-in—"

"You guys can do whatever you want," I interjected, "just don't do a side-by-side."

As it turned out, some networks announced the verdict by scrolling the news like a stock ticker during President Clinton's address. Others used a split screen or opted to hold the announcement until both speeches were over. While I totally agreed with the guilty verdict that night, I found it an occasion for sorrow, not joy. No one likes to see anyone of such talent who was given so many opportunities in life, take a path of violence and hate. It is tragic, but to pay no price for that evil would have been an even greater injustice.

Even without worrying about O. J., I had another image problem on my hands that evening. A couple of days earlier, a *Washington Post* reporter had interviewed me for a personality profile scheduled to run the morning of the speech.

"Are you aware," the reporter asked, "that some black leaders call you a sellout, an Uncle Tom?" These were incendiary words. As is true of any ethnic group, black people sometimes make exacting distinctions as to who is the true bearer of the group's identity. The reporter's question put me in some fine company—Louis Armstrong, Martin Luther King, Jr., and Ralph Ellison had all been treated to similar disparagement at some time in their lives. But that didn't make the words any easier to swallow.

After two years in Washington and seeing firsthand the politics of personal destruction at work, I should have expected a few shots from my "brothers." But instead of chalking up their insults for what they were—partisan nonsense—I shot back. I replied that I knew some black leaders felt that way, and that's fine. This is America, and they're entitled to their opinions. So am I. And in my opinion, I told the reporter, some so-called leaders out there are "race-hustling poverty pimps" whose careers are based on keeping black people dependent on government handouts.

"They talk a lot about slavery," I continued, "but they're perfectly happy to have just moved us to another plantation. What scares them the most is that black people might break out of that racial groupthink and start thinking for themselves."

It was a strong response to a particularly personal assault on my character and my blackness. A part of me was relieved to have finally gotten that off my chest. But the minister in me knew better. Instead of feeling good, I found myself feeling much like I did when I punched the guy at the YMCA back in Norman. There I was crossing over into the wrong because I was in the right. Still, when I had used the phrase "poverty pimps," I had not connected it to any particular person.

When the *Post* article came out, however, it was linked specifically in print to both Jesse Jackson and Marion Barry. The linkage was solely the reporter's inference, and it was made without my knowledge or consent. Both Jackson and Barry have never been in my corner politically, nor have I been in theirs. Despite similar upbringings, we espouse very different basic philosophies. I try to respect all points of view, even those I strongly disagree with. I've also tried not to take cheap political shots at them over the years. Sadly, many black leaders have not accorded black conservatives the same degree of deference.

Unfortunately, in this instance, I didn't live up to my own standards or Scripture, which tell us to seek out our brother whenever we have a difference and talk about it. When the story broke, whether aimed at Jackson or not, the words themselves were enough to set off a media firestorm that threatened to drown out my real message as the controversy grew.

Fortunately, pressure was nothing new to me. I suppose I should have been as nervous as a turkey on Thanksgiving, but the experience of all those years in the harsh media spotlight of college and pro football gave me a certain steadiness that came into play that night. Just as I had never started a game thinking we would fail, I was confident in my ability to complete the task at hand.

A buddy once described me as the kind of guy who would "go after Moby Dick in a rowboat and take along the tartar sauce." Still as I waited in that room, watching the president give his usual first-rate oratorical performance, I suppose it was natural that for a moment or two I wondered if my friends and I had prayed long enough or hard enough for me.

It was at one of those moments that Ed Goeas, an opinion pollster, walked into the room and saw me sitting there with my "game face" on and said, "Don't worry, J. C., nobody's going to be watching you anyway."

When the cue was finally given, I turned to the camera and said, "I don't intend to take a lot of your time. It's late and there's been a lot of talk already this evening. But I want to tell you a little bit about where I'm from."

Then, I went on to paint a picture of my home district for the millions of Americans watching and listening that night because I wanted them to know that the district I represent is as diverse as America itself. It was important they understand how much we all shared despite our differences.

I also wanted to offer them a serious and hopeful vision, articulated in clear language. This speech was my chance to tell people not only what I believe, but what the Republican Party believes and what we are working for. And I wanted them to put Washington into proper perspective; to understand that the state of the union isn't determined at 1600 Pennsylvania Avenue any more than it is on Capitol Hill, where I was speaking that evening. It is the people, all of us, who decide the real state of the union.

I told them I had seen the state of the union just the week before, in Marlow, Oklahoma, where I sat with a group of people at the local elementary school. We ate beef brisket and baked beans, and everyone felt pretty good about how America was doing when the Chamber of Commerce recognized the McCarleys as the Farm Family of the Year. Such communities are where the strength of this country is found and where the state of our union should be measured—back home, where people live good lives and make time for both faith and family. The strength of this country is not on Wall Street but on Main Street; not in big business but in small businesses; not in the marbled halls of Washington, D.C., but in city halls all across this land.

I outlined three key actions that we Republicans intended to take in the coming year to refocus the national lens away from Washington back toward the towns and cities and the people of America.

First, we wanted to make more room for our traditional and spiritual values in our public policies in order to find new models to solve the old problems that have plagued us for generations—education, poverty, crime, and health care.

Second, the Republican Party was determined to enact a balanced budget. We were paying billions of dollars in interest every year that could have been better spent strengthening Medicare; finding a cure for cancer, fighting drugs and crime, or reforming our children's education. It was long past time to take that step.

Finally, I talked about the racial problems that continue to plague our nation. This country must be a place where everyone feels a part of the American dream, I said, and it doesn't happen by dividing us by race or class, by reducing our values to the lowest common denominator, or by using the politics of fear. I told the audience that we must dare to take responsibility for old hatreds and fears and ask God to heal us from within.

Then, 20 minutes later, it was over. I'd done my best, and I felt a wonderful sense of relief. I could start breathing normally again. My staff was standing on the sidelines, happy with the results and also glad it was over. I'd hoped I had touched some hearts and changed some minds about who I was and what the Republican Party really stood for.

I found out later that more people watched the response to the State of the Union Address that night than ever had before in history. I'd love to be able to say that the thought of listening to J. C. Watts, Jr., had them glued to their seats. But I have to be honest and say that the O. J. verdict coming when it did might have given our ratings at least a small boost. I'll never really know, but the remote control is a mighty weapon, just like the ballot box, and the millions listening could have zapped me for HBO or gone to bed. Yet they stayed with the broadcast even at that late hour.

People will never know how much I appreciated their willingness to stick with me that night and the messages that arrived afterward by phone, fax, e-mail, telegram, and letter. They came from liberals, conservatives, the old, the young, women, men, city dwellers, country folk, and pretty much every ethnic group and geographic region imaginable. The vast majority were positive, and, like the Energizer Bunny, they just kept coming and coming.

It was the tonic I needed because as much as I enjoyed giving the response, I didn't like some of the "getting" I got. Not everyone liked the speech. Some thought it "too religious." Some thought it wasn't "Republican enough," that I had dropped the partisan ball. Still others complained that it wasn't "conservative enough." And some of these criticisms were from my so-called friends.

Gary Bauer, the head of the Family Research Council and a man with presidential aspirations at the time, covered the media and key opinion leaders in Washington and around the country with a flurry of faxes condemning my speech because I didn't specifically mention the right-to-life issue. That opened my eyes to the way many people operate in the political system. In spite of everything I'd said in the past about abortion, in spite of the many pro-life votes I'd cast, because I hadn't directly addressed the right-to-life issue in the speech, Bauer apparently felt justified vilifying me in the press and with his many supporters.

His response illustrates just how out of kilter the system can be. My speech was not about issues. It was values-oriented. And that's why so many other people from all across this country responded so favorably. It was intended to demonstrate how much common ground we share, regardless of party, when we think about the kind of nation we want our children to inherit.

Someone once asked me which was tougher, playing football or being a member of Congress.

"That's easy," I told him. "Being in Congress. With football, at least you know you're getting blitzed by the other side!"

Bauer's reaction taught me something new. Independent spirits and political ideologues are seldom compatible. The Jesse Jacksons, Julian Bonds, and the Gary Bauers of the world share very little when it comes to political philosophy, but, ironically, they march in lockstep when it comes to demanding political purity as the price of admission to their respective "groups."

I had known for a long time just how strong the group identity issue was in the black community, especially with national black leaders. They believe all black people should think and act alike. If you don't, you're accused of forgetting where you came from or you become the target of the kind of invective that was leveled at me for giving the State of the Union response. What I hadn't understood was the importance of group identity to conservative and even Christian leaders involved in politics.

Before I had time to let Bauer's attack sink in, I found myself in more hot water with my fellow conservatives. I decided to apologize for my overheated language and affirmed that my "poverty pimps" statement was not aimed specifically at Jesse Jackson or the mayor of Washington. I didn't back away from the substance behind my original statement—I do feel some people use race and poverty to benefit themselves—but I regretted the inflammatory, inappropriate language I'd used to express it.

As a congressman, a minister, and a role model for both black and white kids, I had a responsibility to maintain a certain standard. In making those kinds of comments, I hadn't met that standard, and I didn't intend to let my mistake stand. Apologies are golden. Sometimes, they're the only way to reconcile a brotherly disagreement before it turns into all-out war.

Paul Weyrich and a few other influential conservatives immediately took me to task for "backing down." I was stunned. What kind of society do we live in if you're wrong for apologizing? If the tenor of my words caused hurt, even if the substance of those words was right, then why not apologize if it will help bring a tone of civility back to the discussion?

Criticizing me for not being "Religious Right" enough or conservative enough is just as ridiculous as hearing from some people that I'm not black enough. Having found myself caught in the crossfire between the extreme right *and* the extreme left in the same week left me questioning my own place in the political arena. I spent some long hours pondering the criticisms of both sides, but in the end, it seemed to me to boil down to some plain facts.

I'm an ordained minister. I'm a proud Republican who has so far won every election I've entered. In addition, I've had a conservative voting record I would never walk away from, but most important, I have the dictates of my own conscience and the strength of my own convictions. I don't need the input of any lobbying organization to help me understand what it means to be a Christian or a conservative. And I've certainly been called enough derogatory names and fought exclusion enough to have no doubts about my identity as a black man in America.

While my background in athletics was a great training ground for life in many ways, it could never have prepared me for the willingness of fellow team members to sacrifice a collegial working relationship for a short-term media gain. In sports, someone who is wearing the same uniform as you is your teammate. You may have disagreements. One may gripe about the other's performance inside the locker room. But when the game is over, teammates don't undercut each other. Too much is at stake. The whole team's destiny is bound up in winning, and great team players sacrifice their individual glory whenever necessary to help the team win.

Think of Michael Jordan, arguably the greatest basketball player the world has ever seen, passing up the game-winning shot, which was also the series-winning shot, in the 1998 NBA finals. Steve Kerr, a player with tremendous long-range accuracy, had a cleaner shot at the basket, so Jordan passed off and Kerr became the hero. That's consummate team play.

Politics as commonly practiced today doesn't leave much room for forgiveness and reconciliation. We've really lost the human element. When you play a football game, the object is to win, and there's no question about it. But eventually, 60 minutes run off the clock, then you stop and understand that the people you've been competing against are human. They are somebody's son or brother. If I've learned anything over the past eight years in Washington, it is that politics shouldn't be any different.

We should do a better job of reminding ourselves of that human fact. Party lines are not always the straightest lines to the truth. Neither are the lines of a stereotypical ethnic identity. Or, for that matter, the theorizing of a conservative or liberal think tank.

I believe you can be a productive, supportive member of a group yet retain your independence on specific issues, particularly issues of conscience. Within the Republican Party, we have differences of opinion on abortion, the environment, stem-cell research, health care, and even taxes. This diversity of viewpoints doesn't make Denny Hastert's job an easy one, but all members have their own individual constituencies to serve and their own deeply felt beliefs and principles upon which they base decisions. And every member must live with those decisions.

I'm a pretty conservative guy on most issues, so in 1996 and 1997, when Congressmen Gary Franks and Charles Canady began a legislative effort in the House to limit or outright ban affirmative action in the federal government, most of my colleagues expected me to jump on the anti-affirmative action bandwagon. But I held back. I talked with Gary, a terrifically talented public servant, about his bill to ban the use of racial preferences in government contracting. I'm against race- or gender-based preferences, but I also felt Republicans should invest at least as much energy proving we are inclusive as we do weaning ourselves from affirmative action. The debate surrounding this issue is both intense and divisive. I've had more than one member of Congress from both sides of the aisle tell me privately, "I wish we didn't have to vote on this."

But, of course, we do. It's our job to make the tough calls based on the clearest conscience and the best wisdom we have as individuals. Affirmative action is a little like the professional football draft. The NFL awards its number one draft choices to the lowest-ranked team in the league. It doesn't do this out of compassion or guilt. It's done for mutual survival. The commissioner and the owners are smart businesspeople first and foremost. They understand that a league can only be as strong as its weakest team.

In much the same way, there are many people near the bottom in our society who, given the right breaks, given clear access to incentives, could turn their situations around. And when they do so, it works to the betterment of all of us. The NFL does not make the number one draft choice an entitlement, however. Just because you were an expansion team in 1991 or the lowest team in the league in 1997, doesn't mean you will get the number one draft choice forever.

The same goes for affirmative action. I have always believed a better criterion for extending this kind of help would be to base it on economic need alone, not race or gender. Economic need isn't confined to one gender or ethnicity. Race-based solutions, however, feed on the notion that membership in a certain race is a handicap, a sure cause of underperformance. And that's pure nonsense. Worse, it encourages some ethnic groups to wrap themselves in what I call the cocoon of victimology.

But in 1996 and 1997, with California's Proposition 209 providing momentum, Republicans wanted to take an ax to federal affirmative action programs, and some thought I should be out front leading the charge. While I partly agreed with the concept, I felt deep down that the time was not right to move on this controversial issue. As Republicans, we had done little to prepare America, much less black America, for what would no doubt be characterized by the media and black leaders as a major rollback of civil rights for minorities. We needed some thoughtful discussion within the party before acting rashly.

I asked Gary to pull the bill, to wait for a better moment and give us time as a party to really think this problem through. But in the heyday of Proposition 209, the California referendum to end the use of racial preferences for entrance into the state university system, conservatives in Washington were clamoring for similar action at the national level. Gary felt just as strongly that this was the right direction to take as I felt it would be a huge mistake. He decided to move forward.

This left me no choice but to go to Newt. I rarely sat down one-on-one with the Speaker. Trailing him around the Capitol was like trying to track a tornado. Getting a few quiet minutes was never easy either, so when I requested a meeting, I wanted it to be critically important. From my perspective, this was. Affirmative action had been a red flag in front of conservative bulls for many years, but as conservative as I was, on this issue I had to part company. As one of only two black Republicans in Congress, I had felt the sting of racism; I had attended a segregated school, sat in the "black" section of the movie theater, and been the target of the "N" word more times than I cared to remember. I had paid my dues plenty for this meeting and for the right to my independent view.

Newt and I sat down together in his office. I talked to him from my heart. I told him I feared a federal ban on affirmative action would send the wrong signal. It would say to black Americans, "We don't believe racism exists."

I went on to say, "Newt, I don't think we've laid the foundation to communicate with the American people what this issue is all about."

I told him we needed to go slow in a way that proved we understood that the playing field isn't level yet for some minorities and demonstrate that we are determined to keep pursuing equality. I said that I knew people on both sides of the aisle, white and black, liberal and conservative, who fully understood that affirmative action is flawed. But most would also encourage us to proceed carefully while evolving something better.

We agreed that neither of us liked racial preferences, but I said, we have nothing to replace them with at this point, to give people some comfort that we understood their concerns—no plan B.

Finally, I said, "Maybe I should see this as a political issue. Maybe I should think in terms of Democrat versus Republican. Maybe if I could stop leading with my heart, I could support anti–affirmative action legislation, but deep down, I know it's wrong for the party to move at this time."

I'll never forget that moment. I didn't know what he would say or how he would respond to what amounted to conservative heresy. Then the Speaker leaned over and put his hand on my arm.

"That's why I like having you around, J. C." Newt smiled. "Don't ever stop listening to your heart." Mr. Conservative had given me the green light. Yes, this party is big enough to handle an independent cuss like me.

Newt went to Congressman Franks, who withdrew his legislation—not happily but with dignity and class. For me, one of the biggest disappointments of the 1996 election was the defeat of the only other black Republican in Congress, Gary Franks.

But the affirmative action debate wasn't over. Charles Canady, a Florida congressman at the time, introduced a broader legislative package to end racial preferences. This time supporters brought in the big gun, anti–affirmative action proponent Ward Connerly, to try and convince me that Proposition 209, which California voters had passed in November, signaled a turning point in the affirmative action debate. "It's time to get on board" was the message. A large group of us, including Newt, met late one afternoon to discuss the issue. Connerly insisted that affirmative action was antidemocratic and worse. He was opposed to my position. I, on the other hand, basically repeated what I had told Newt a few months before. Connerly wasn't buying it. For him, the defeat of affirmative action was a matter of principle, but he refused to understand that this was a matter of principle for me as well. The discussion heated up, and finally I let fly.

"Affirmative action isn't the problem," I said. "Lousy education for black kids is the problem. Until you fix these schools, don't talk to me about equal opportunity. I know better."

But I wasn't finished. I told them all a story. Not long before, I had been home in Oklahoma driving my S-10 Blazer to the dry cleaners one morning when not one cop but six pulled me over in broad daylight. They checked my driver's license and registration. I hadn't been speeding. My license plates were current, and I wasn't weaving.

"I was stopped for one reason and one reason only," I said. "Because I was a black man.

"I could be a member of Congress, a Republican leader, the man who gave the response to the State of the Union address, and one of the most well known football players in the state of Oklahoma, but in the eyes of those policemen, I was only one thing—black." The room went quiet.

Racism isn't dead in this country, and if some in the room hadn't experienced it personally, that didn't mean it wasn't real. Moreover, as I sat there, I thought how ironic it was that this group, black and white, that so desperately wanted to take race out of the picture, came to me for support, not because of my oratorical skills or my ability to persuade, or my legislative influence, but because I was black.

When the meeting ended, neither side had moved, but I felt a whole lot better having said my piece. My independence on this issue had put me slightly in left field, but I knew Newt understood my position and I knew it put him in a difficult position, too. He was already under fire from conservatives for his decision to delay a tax cut vote and for the lagging pace of the conservative agenda. If he had decided to pull back publicly on affirmative action, the coup that came a few months later might have come earlier.

When Canady's bill came up for a vote in committee, two conservative members unexpectedly changed their votes, and the effort to end affirmative action was dead. I don't know if Newt had anything to do with those switchers, but I do know that he's never gotten the credit, particularly from the media and the Black Caucus, for his efforts to take a reasoned approach to this very difficult issue.

For all Republicans but especially for Newt, 1997 was a rollercoaster year in more ways than one. That summer, we had finally passed the Balanced Budget Act. After a two-and-a-half-year feud with President Clinton, we finally were able to force him to accept two long-term Republican goals—a balanced budget and tax relief. This was the reason I had run for Congress—to make this kind of history. We not only provided tax relief for families but saved Medicare from bankruptcy and began the process of restoring power, money, and influence back to the states.

The balanced-budget agreement couldn't have come at a better time. For months, Newt had been stymied by accusations of ethics violations. Finally, he paid a fine to put an end to the media feeding frenzy. But by now, Republican poll numbers were trending down, and a number of members had lost confidence in Newt's ability to lead. Just 18 months before, Republicans were riding high, coming off their historic victory that delivered control of the House. "Bill Clinton is toast" was the mantra of the time.

How wrong we were. Like a clever cat, President Clinton seemed to have nine lives; and before we knew it, he was back stronger than ever, challenging Newt and the Republicans at every turn. In the 1996 elections, Republicans had lost seats, which raised the first doubts about Newt in the minds of some members. By the summer of 1997, Newt was in real trouble and pundits all over town were predicting his demise.

One day, a member approached me just off the House floor and said, "We're going to bring Gingrich down. You don't have to be involved in it. I just want you to know." This was one of the more prominent members in the House, and it was clear that his statement was no idle remark. Others were also told, and Chris Shays of Connecticut alerted the Speaker's office. The Gingrich forces jumped into action. Newt and company contacted all the committee chairmen to shore up their support. Once the chairmen were on board, the coup died a fast if not painless death, but the damage to Newt had been done.

I think there were probably thirty to thirty-five members who saw everything Newt did with a jaundiced eye. I was not one of them. I considered Newt not only the House leader but a friend as well. By overcoming the coup and passing the Balanced Budget Act, Newt was able to consolidate his power base once again as we went into the fall. I had hoped that restraint might be the watchword for the spending bills that were now coming to the floor, but I was to be disappointed again. After the disastrous budget debacle of the previous year, some believed just avoiding a government shutdown was a moral victory. Given how much of the people's money we spent that fall, however, I'd call it a pyrrhic victory.

Election year was upon us before we knew it. We all returned to Washington after the Christmas recess ready to tackle a broad set of issues. Republican poll numbers were up, and most of us hoped that Newt's problems were now over. From my point of view, it was high time we focused on key issues, like education, community renewal, health care, and a much-needed focus on defense. But for all our good intentions, this was not going to be a year of major achievements. In January 1998, Monica hit town like an out-of-season hurricane that no one was prepared for. I had been in Congress for 3 years at that point and was still a little wet behind the ears by Washington standards, but I don't think even the wisest veterans could have predicted the impact this scandal would have on Capitol Hill. Partisan politics will always be a part of the political process, and that's as it should be. But it seemed as if the Clinton presidency sharpened the political edges. It wasn't just politics anymore; it had become personal. Newt's problems were living proof of that.

My entire tenure in Congress had been served under a Democrat president, so I had no basis for comparison. But many of my colleagues complained about the harsh change in tone over the past 6 years. Democrats blamed Newt. Republicans blamed Clinton. The truth probably lay somewhere in the middle where regular folks are found.

But I will say few people would argue with the notion that the Clinton people played hardball politics at its hardest, and to be blunt, Republicans just weren't as good at it. It cost us politically; but from my point of view, that is not an attribute to which we should be aspiring.

I remember seeing a feature story once on ESPN profiling the great NBA forward David Robinson. The interviewer asked him about his reputation as a good guy. "People think you're too nice," the reporter charged, as if being nice was something akin to a criminal offense. I thought to myself, "Which would you rather have people say about you—he's a nice guy or he's a horse's rear?"

I've been accused of the same "problem" a time or two, but the no-holds-barred kind of partisanship that had captured Capitol Hill by the time the Monica Lewinsky scandal hit had become destructive. I'm not suggesting we should have looked the other way, labeled the president's perjury and lies to the American people "a personal matter," and moved on. But the scandal, combined with an already poisoned atmosphere, made it difficult for both Democrats and Republicans to do their jobs.

L'affaire Lewinsky, as Fox anchor Tony Snow called it, was like an all-consuming virus that infected everything we did. By fall, Republicans found themselves in the unpopular position of considering impeachment, while the president bashed us over education reform, something Americans wanted passed. He offered 100,000 new teachers, and we were offering, in many Americans' view, a degrading romp through President Clinton's personal life. We were caught between a rock and a hard place: ignore our sworn obligation to uphold the Constitution or put politics before duty.

Despite sinking poll numbers, Newt believed electoral history was on our side. He told skeptics that the American people would reject the party of scandal, and Republicans would see major gains in the fall election. He was as convincing on that point as he had ever been on anything.

I didn't have time to worry about the national picture, however. I was toughing out a mini-scandal in my own race. Just when I thought partisan politics couldn't get any lower, it did. Somehow, my opponent had gotten hold of the FBI tape of my conversation years earlier with Bill Anderson, the lobbyist who got himself into trouble.

The recording should have been under lock and key. If taken out of context, which it was, the tape was damning. Fortunately, the *New York Times* reprinted the entire conversation, including my clear statement to Anderson that I would not trade favors for contributions. How my opponent got his hands on a tape from a secret government investigation remains an open question. But if people in a position of trust in our federal government handed the information over, which they did, they were guilty not only of illegal actions but of breaking the oath to defend and protect the Constitution of the United States. Whoever it was failed in that mission. I was reelected with my largest margin ever—62 percent—and my experience in that campaign only reinforced my belief that government as it exists today is too big, too intrusive, and too willing to play the role of Big Brother.

The news back in Washington, however, wasn't good. Despite Newt's confidence that we would win 10 to 30 seats, Republicans took a bath in the election. Traditionally, the party that controls the White House almost always loses seats in an off-year election. That's what Newt, a dyed-in-the-wool historian, and many others were counting on to bring us through. They couldn't have been more wrong. For the first time since 1934, the party outside the White House—the Republicans—lost seats in the midterm election. We made history all right—the wrong kind.

We shouldn't have been surprised. The polling data hadn't looked good for months, and then, just a few weeks before the election, Republicans made what I believe was a huge strategic error. In an effort to tie Democrats to the Clinton scandal, which had now reached historic proportions, the national Republican Party went on the air with a multimillion-dollar

Lewinsky ad campaign in key House districts around the country. Even though ad testing focus groups had reacted negatively to the spots, the decision was made to go ahead. It was a huge mistake.

I don't think we would have lost seats in that election if we had campaigned on our strengths—on what we stood for, what we'd accomplished, and what we proposed to do. Instead, we tried too hard to invoke the public's disgust over the travails and peccadilloes of President Clinton. Lord knows, there were plenty of them, but we made a strategic mistake by making Bill Clinton the focus of the election.

Like most members, I didn't have a say in that decision and really didn't know the depth of what went on until after the election. But I guarantee you, if someone had tried to run commercials in my district about the Lewinsky scandal, I would have balked. Republicans had a great story to tell, we just needed to tell it better. I believed then and now that we had accomplished too much and stood *for* too much to be running *against* something or someone.

I really believe people want to know whether you share their concerns and values and what you intend to do to help solve their problems. Republicans had balanced the budget for the first time in 40 years without using the Social Security surplus. We'd passed tax relief, reformed welfare, and strengthened Medicare. Interest rates and unemployment were low, while the economy remained strong. Yet, in district after district we were talking about President Clinton!

Once the dust had settled, Republicans kept control of the House by the skin of our collective teeth. The long knives were out for Newt, however. Just as he had gotten the lion's share of the credit for the 1994 victory, now he was to get the majority of the blame for the 1998 defeat. Much as I liked and admired Newt, when leaders face disaster on their watch, the responsibility falls on them, and Newt was no different. In the days after the election, everything from a larger black turnout inspired by Democrat-funded hate radio to the Republican Senate's refusal to take up the House tax cut got some of the blame for our poor showing. But the majority of the fingers were pointed directly at Newt, and he accepted responsibility. I remember him telling me in a conversation just after the election that he finally understood how a head coach must feel.

"We won," he said with resignation in his voice, "but not with enough points on the board." I thought of my years at OU when meeting expectations was almost as important as the victory itself, and I probably understood better than most what Newt was feeling in those days after the election.

The Speaker's call had come on the following Friday morning. I was still home in Oklahoma, enjoying the first peace and quiet I'd had in months. By now, Bob Livingston, the powerful chairman of the Appropriations Committee, was out publicly campaigning for the Speakership and so Newt got right to the point by asking for my support. I told him, "Newt, I'm your friend. I like you. You've been very good to me. But as a friend I need to tell you that that little half-cocked coup attempt was probably a mile wide but not very deep. Your current problems are much deeper."

I passed on what I had picked up over the past couple of days talking to other members about the election. It wasn't good news.

"Man, I hate to see you put your family and yourself through this," I said, "because I know if you stay, it's going to be ugly." We talked for nearly a half hour. I said, "Newt, I've had my own fans boo me. I know it's no fun."

He asked, "How do you stop the boos?"

I said, "Barry Switzer used to tell me all the time, 'The way you stop the boos is you go out and produce.'"

Later that night, Pam Pryor, my chief of staff, called to tell me she'd just gotten word that Newt was going to resign. It didn't surprise me at all. I had sensed earlier in the day that he was thinking seriously about it. He had been more reflective than I had ever seen him. Newt was a man of action. His style was to go at things, attack, and launch the initiative. He didn't often sit around and ponder.

And Newt was always a fighter and a veteran of plenty of tough campaigns. He was the man who brought down former Speaker Jim Wright. He was and still is a fascinating and brilliant man, but many of us believed that Newt's talent was in his ability to buck the system, not manage it; to rock the status quo, not become a part of it; to lead change, not govern day to day.

There are few cases in history where the revolutionary outlives the revolution. In the end, Newt was a casualty of his own success and failure. In the end, he announced without rancor or recrimination that he would hand over the gavel and head for private life. Within 24 hours, the man who had pushed him toward resignation, Bob Livingston, would announce his intention to resign from the Congress as well, and House Republicans, reeling from a major defeat, were thrown into a bitter internecine battle. Who would lead the next Congress? With the Republican margin now razor thin, what kind of man or woman should take the helm? What about the rest of the leadership positions? What would all this mean for the impeachment sure to come in the months ahead?

No one knew the answers, but the maneuvering for power had begun. Meanwhile, I was sitting out in Oklahoma looking forward to taking some time off. Frankie and the kids were overdue and ready for my undivided attention. It was *their turn,* but some folks in Washington had other ideas. They decided I might be one of the answers to their leadership problems. It sure wasn't on my agenda. I had nothing more serious planned than sleeping in for a change and spending some quality time with the family.

But on the day Newt resigned, the phone calls began as one member after another phoned to urge me run for the Republican Conference chairmanship. I'd already heard some rumblings about a shakeup in the Republican congressional leadership, but running myself was honestly the furthest thing from my mind.

I made a few calls, talked to my staff, and then sat down with Frankie. Once again, she was my rock. She understood the honor of a leadership position and was willing to put up with one more campaign that year. By the next day, I was off and running, but at least we knew this craziness could only last until November 18—the date Republicans would meet to organize the House and elect its leaders—about two weeks away.

This campaign, however, turned out to be a whole different breed of cat. When it's Republican versus Democrat, that's one thing. But intra-party politics can sometimes be just as brutal. As soon as we made the decision to go after a leadership post, we put together a SWAT team of folks to lobby for my election. All of us worked our tails off for the next 12 days. The magic number we had to get this time, however, was slightly less than my congressional election—only 112 votes.

Our target was the 223 Republicans that would make up the next Congress. I tried to call every one of them. Members have only one vote, and they're pretty particular about how they cast it. People running for a leadership post should expect to find themselves scrutinized and on the receiving end of some pretty good questions. It's a generally fair process, but it has its tough moments. At one point, some overeager staffers on the other side dredged up the old Oklahoma Corporation Commission scandal to tarnish my reputation in the eyes of my colleagues. As unethical as it was and despite the fact that I had been cleared of all charges, some people gave the rumors credence.

Much to my surprise, Steve Horn, a moderate member from California, came to my defense. Steve is affectionately known as "Mr. Oversight," and no one in the Congress gets to the bottom of government waste, fraud, and abuse better than he does. While the two of us certainly knew one another, we didn't run in the same circles, but Steve took it upon himself to get to the truth of the rumors that kept swirling around my candidacy. When he discovered that there was nothing to the charges, he got the word out, telling members, "Vote against J. C. if you want to, but don't vote against him because of these rumors. They're simply not true."

Even though Steve and I had our philosophical differences then and still do, I have never forgotten his willingness to put truth above ideology, and I will always be grateful.

Even without the scandal rumors, convincing members of the wisdom of my candidacy was at times painstaking and occasionally painful. Just one example was Jim Talent, my partner in the American Community Renewal effort, who had the unpleasant chore of telling me he had committed to voting for my opponent, the current conference chairman, John Boehner. I told Jim not to worry, I respected his choice, and it wouldn't affect our partnership or friendship.

The fact that I would be the first African American of our party to hold a leadership position wasn't an issue in these conversations, but I would be less than honest to think my race had nothing to do with my election, at least in the minds of a few. I don't believe in racial preferences, and I hope no one voted for me just because I was black. The color of my skin was not a qualification for the job; but, perhaps, it was an added benefit. If it was one of the factors that members considered because they wanted to reach out to the African American community, to send a signal that our door is open, so be it. I take pride that my party chose to put a person of color in the House leadership.

After all my calls were done, the strategy completed, I predicted I would get 123 votes. We ended up with 121. On November 18, my forty-first birthday, I found myself a part of the Republican leadership after only 4 years in the House. Not long after I was elected, my friend Pat Adams called me at the office and asked how it felt to have won the post. I told him my mental state was best summed up by "I caught the ball—now where do I run?"

Gerald Ford, Jack Kemp, Dick Cheney, and Dick Armey are among the people who have held the conference chairmanship in the past. Traditionally, it's been a kind of free-form assignment. You pretty much make of it what you will. At a minimum, you are responsible for a conference meeting of all Republican members usually held at least once a week. I've chaired those meetings, which are called to discuss policy and politics. Leaders from President Bush to Speaker Hastert have attended conference sessions, which give members an opportunity to get things off their chest, plan legislative strategy, and get information on key issues and the Republican message.

Because of my reputation as a speaker, I think most people expected me to use the post to become a party spokesperson. That is an important part of the job; but as I took over the helm of the conference, my first concern was to build a system to communicate and provide members with the tools to get our message out across the country. I had been disappointed in our election strategy. All the good things Republicans had accomplished got lost in the focus on President Clinton's shortcomings.

I also knew I wanted to use this position to reach out to people whose view of the Republican Party had hardened into mistrust and dislike. Scripture says we should be fishers of men. As Larry Thompson, a pastor friend of mine, would say, "You catch more fish with a net than a gun, and I've never seen a fishing pole that would shoot." What he's talking about is the best way to reach out to people to bring them into the fold, and it applies every bit as much to a political party as a church.

Too many Republicans leave their nets at home when looking for converts. The message people hear from them isn't one of diversity or tolerance for differing points of views.

They hear, "If you don't believe what we believe on every issue, you don't belong in our party."

"If you don't fish the way we fish, you aren't one of us."

That's the wrong message to send for a party that wants to grow. When you fish with a net, you're also likely to catch more than one kind of fish. As a party, we needed to do a better job of recognizing that this is a diverse land growing more diverse every day as people from all over the world have become Americans and many more clamor to come to this great country. We need to reach out to these people, make sure our welcome mat is also in Spanish and Arabic, in Asian and African languages.

Often when Republicans propose legislation to help people, they defeat themselves by speaking in cold economic terms that make us seem heartless, or at least bloodless—what I call "Washington speak." Politics is the competition of differing philosophies for the hearts and minds of voters. In this age of 24-hour news, we've got to be smarter about how we present our ideas and ourselves to the American people.

Many Republicans who are in safe, conservative districts often approach issues this way: They'll tell you, "I'm right, and I'm principled. This is the principled thing to do, and by golly, I'm going to do it." I suspect my votes haven't been much different from theirs on most issues— pretty conservative. What has separated us is how we communicate those views in terms people can relate to, in terms of benefits, not policies. I made one of my responsibilities as conference chairman the mission of saying to my Republican colleagues, "Look, I think you ought to also consider *how* you talk about issues. Consider what people are hearing, not just what you are saying."

In our churches, we call it outreach. Our job is to communicate our faith to newcomers or outsiders in a way that conveys the truth of our beliefs and at the same time welcomes them to our group. If you visit a church for the first time on a Sunday morning and see a congregation that looks like they've all been sucking lemons and no one extends a hand in friendship, you're not likely to come back. A smart pastor tells his flock, "Smile. Be friendly. Let people know you're glad they came to our church today."

At times, I think some Republicans are afraid of being nice; maybe they're afraid they'll be accused of catering to special interests. But all politics is really special interests. Pro-life groups are special interests and so are their pro-choice counterparts. Antitax groups are just as much a special interest as the trial lawyers or labor unions. We need to remember that politics is all about people, not programs. We shouldn't want to take the humanness out of the political arena.

That's how I saw my chairmanship—to give Republican members the tools to open lines of communication with all the diverse communities of our nation. Luckily, my colleagues had elected a Speaker who understood the power of teamwork. While I didn't know Denny Hastert well, I knew he was a former teacher and ex-coach. That said to me Denny was probably a smart, focused, dependable person with a good heart for leadership. It also told me he was probably good with people, too. I was right. Denny is a solid leader who keeps his ego in check. Sometimes, people criticize him for being less in the public spotlight than Newt, but that's an unfair comparison.

Denny Hastert is a workhorse. He stays focused on the job and doesn't get distracted by the media hype that usually swirls around the Speaker's office. For members, he's always a good listener. On the occasions I've gone to him on matters of principle, he's heard what I've had to say. He's also given me the independence I've needed both as conference chairman and as J. C. Watts, Jr. I always felt that while I may be in leadership, and that demands a certain degree of allegiance, I cannot abandon my own beliefs and priorities simply for party unity any more than any other member can or should.

Last year, I called for House Republicans to hold hearings on the high gas and electric prices that were plaguing much of the country. Too often Republicans beat up on big government but don't want to take on big business. I didn't ask for the hearings, however, to bash the oil companies and score partisan points as our Democrat friends in the Senate had planned. I wanted a hearing because I thought the American people deserved what I called "price accountability." They were the ones paying the electric bills and the prices at the pump. It seemed to me they ought to get an explanation for why prices can skyrocket in a matter of a few days.

I suggested bringing representatives from the oil companies, who were getting pilloried in the press, into Washington for a thoughtful discussion on pricing. My suggestion wasn't meant to be threatening, but I felt it was time to cut through all the questions and get some answers. Dick Armey didn't cotton to my idea and said so to a group of reporters, who promptly turned our disagreement into a full-blown feud. Once Dick understood what I meant by price accountability, he apologized for his initial criticism and backed the idea.

I'm always amused by the liberal media's insatiable appetite for what they see as either "black-on-black crime"—J. C. disagreeing with the Congressional Black Caucus—or Republicans beating up on Republicans. When it comes to press coverage, Republicans never suffer from media attention deficit disorder as long as we're being "disagreeable."

If I'm at odds with Jesse Jackson or Gary Bauer or Dick Armey, I get plenty of coverage. But it's not just me. Most Republicans get this special treatment by the media. Late last winter during an MTV appearance, Colin Powell answered in a straightforward manner a question on the use of condoms posed by a young lady. Whether you agree or disagree with the secretary's views, that Colin Powell might have slightly more moderate views on certain issues is not exactly breaking news. But the media went into a veritable frenzy, trying to paint a simple answer into a monumental split in administration unity.

The fact is, Republican members of the administration, party leaders, and elected representatives all have their differences of opinion from time to time. So do the Democrats. We've seen it with the split in the Senate this past year over an economic stimulus package and a national energy policy, but the media never seems to have much of an appetite for covering divisions in the Democrat ranks.

I've learned to take our media spotlight in stride, but what I haven't been able to get used to is the continuing partisanship that neither party seems capable of overcoming despite President Bush's best intentions. No amount of partisanship, though, could ever match the firestorm of impeachment. I've faced many difficult votes during my time in Washington and taken a lot of criticism for some. But the most heart-wrenching vote that I've had to cast was the vote to impeach President Clinton in January 1999. Some might also say it was the most important vote. I'd have to disagree. A balanced budget, a stronger national defense, welfare reform, Medicare solvency, and education reform—my votes on these issues may have been less dramatic, but in the long run they will have far more impact than impeachment on the lives of most Americans, with the possible exception of the nation's historians.

While I don't want to minimize the difficulty of my impeachment vote or its historic nature, we also need to keep it in context. The American people certainly did. The vote by the Senate not to convict President Clinton on the articles of impeachment may have ended the legal proceedings, but the fallout from this terrible episode in our nation's history was far from over. The bitterness and rancor that marked the proceedings, and in fact the whole Lewinsky affair, left a scar on the body politic of our nation that won't easily be removed. That is at least part of the reason why the 2000 election was so close. The impeachment of President Clinton divided our nation like nothing since the Civil War. I, for one, hope this country will never have to face the impeachment of a president again.

None of us are perfect human beings. We all make mistakes. But I could not in good conscience look the other way when it came to President Clinton's actions. I could not simply let his perjury go as nothing more than "lying about a personal peccadillo," as his Democrat supporters on the Hill claimed.

One of the most destructive cultural changes of the past 20 years has been this country's slide into moral relativism. Too many people, especially our young people, are uncomfortable acknowledging that there is right and wrong, good and evil, in this world. It's easier to live in a nonjudgmental environment where abortion has become a method of birth control, drugs are a lifestyle choice, profanity and hate in music cloaks itself in the banner of "free speech," and violence is excused by virtue of poverty.

The moral climate in which we find ourselves today discounts character as an antiquated value better suited to nineteenth-century novels than twenty-first-century America. I disagree. I think character in our presidents matters a great deal. But when it came to impeachment, we were not voting on whether Bill Clinton was a man of good character. Like a grand jury considering a potential indictment, we were asked to judge President Clinton on the merits of the legal case.

Some members saw his perjury as a crime, plain and simple. Others felt it was also a violation of his presidential oath. Still others believed no person should be above the law in this country.

I saw it through my own prism. By lying under oath in a sexual harassment case, Mr. Clinton had exceeded the bounds of common perjury to commit a violation of a person's civil rights. The issue wasn't whether Paula Jones was telling the truth or whether President Clinton did what a lot of men do when caught in such a sordid situation—try to lie their way out of it. The issue was whether Paula Jones, like every other American citizen, had the right to a fair, impartial hearing of her case.

For me, it was a clear case of the denial of an individual's civil rights. I pondered the difference between a case of sexual harassment and a KKK member hauled before a court who lies under oath about racial harassment or, worse, racial violence. For those who opposed impeachment, I wondered what their attitude might be if Paula Jones had been black and had charged Mr. Clinton with racial harassment. I asked myself if most Democrats would have stood by him under those circumstances. Would they have looked the other way when he lied about the facts of the case? Is sexual harassment somehow less onerous to them than racial harassment because it involves an intimate part of one's personal life? It certainly wasn't for most Democrats when the accused was Clarence Thomas. The answer was clearly no on all counts. It seemed to me that my friends on the other side of the aisle had constructed a double standard that was simply not acceptable or explicable in other than partisan terms.

When I thought about impeachment in this context, as a civil rights issue, my decision was made. I believe the president of the United States has a moral and legal responsibility to protect and defend the Constitution and, by extension, the civil rights of every American. As the nation's chief law enforcement officer, President Clinton had clearly failed in his duties by denying Paula Jones her civil rights.

Yet, despite my belief in the correctness of my final decision, I cast my vote in favor of impeachment with true dread, and I prayed I had chosen the right course along with my colleagues. With a few exceptions, I don't think the vast majority of Republicans who voted to impeach the president did it lightly or without great sadness.

In a speech to the 1992 Republican Convention, former Secretary of Labor Lynn Martin warned, "You cannot be one kind of man and another kind of president."

Those are wise words for anyone serving the public trust. I have spent nearly twelve years in public office—four at the Oklahoma Corporation Commission and now eight as a member of Congress. The people of Oklahoma have been good to me and have honored me with their votes and trust. My colleagues have done the same by giving me the opportunity to serve as part of the leadership, and I am grateful for their confidence.

While I am leaving Congress, I do know there is still much work to be done. As I write this book, President George W. Bush has taken the helm and given our country a fresh start for a new century. America is united and steadfast in its determination to win the war against terrorism, while the problems of education, poverty, national security, race, and moral uncertainty cry out for all of us to put aside what divides us and find real solutions to help real people.

13

God Put Us on This Earth, in Part, to Prepare the Way for Those Who Will Come Behind

Every so often we hear people clamor for a change. Let's change the Constitution, change the form of government, change everything for better or worse except to change the only thing that needs changing first: the human heart and our standard of success and human values.

William Boetcker

The morning of September 11, 2001, dawned just like any other day in Washington. Like many fall mornings in the nation's capital, it was warm and sunny. Indian summer had kept the trees full and green on the Mall, and flowers were still blooming along the Potomac, not far from the Pentagon. The morning traffic crawled its way into the city as the Capitol dome gleamed in white relief against a brilliant blue sky. It seemed an ordinary day.

I had left home early to speak to a breakfast meeting of the National Automobile Dealers Association, where I talked about the usual issues and answered a few fairly standard questions. By 9 o'clock, I was back in my office in the Longworth House Office Building when this most ordinary of days became one of the most tragic moments in our nation's history.

I had just returned, unaware that the first plane had already hit the World Trade Center, when I heard Jack Horner, my legislative director, yell, "My God, they flew into it on purpose!"

I could tell from the intensity of his voice that something terrible was happening, and I rushed into his office to find several of my staffers gathered around the television, stunned and unmoving. As the two massive buildings began to burn, our initial shock gave way to total disbelief.

Very few words were spoken. Then, 20 minutes later, the Pentagon was hit. Now the terror came home. It wasn't in New York or a faraway embassy or a ship in a remote port. Terror had come to Washington—just a few miles from where we were standing, watching the devastating pictures of a brutal attack on what is the nerve center of our nation's military might.

It was unimaginable, impossible, yet there it was for the whole world to see. Disbelief was quickly replaced with uncertainty and fear. Could there be more coming? My mind went immediately to the Capitol dome.

I remember saying, "They'll go after the dome next. It's the symbol of freedom for all the world." And I believe today that were it not for the courage and sacrifice of the passengers and crew on United Flight 93, the Capitol might have been destroyed as well on that dark day. We owe them all a great debt of gratitude.

Like the passengers on the doomed planes, our first instinct was to call our families. I told staff to let their parents and spouses know they were safe and then get out of our building, which sits directly across from the Capitol on Independence Avenue. People hurried to gather their belongings, trying to decide where to go and how to get there. Traffic throughout the city was bound to be a nightmare as people rushed to get home.

As anxious as everyone was, there was also a need for comfort and calm, so several of us gathered in my office for a moment of prayer. It was as emotional a moment as I could remember in my seven years in Washington. None of us knew what was happening. Was it over or just beginning? Was it just Washington and New York that were under attack or would scenes of destruction be played out in more cities with more loss of life?

A word of prayer was what we needed, and God gave us the calm to push on. Twenty minutes later, the entire Capitol complex was officially evacuated. As we emerged onto the street that morning, we were met with a kind of controlled chaos as crowds of people, some in a state of shock, some crying but not panicking, streamed from the various House office buildings and the Capitol. It was surreal, as if we had been suddenly transported into a scene from a disaster movie. Minutes later we heard a loud boom. It wasn't a bomb, but the already palpable tension got much worse. Still, no one panicked. Instead, we hurried away from the Capitol and on to a friend's office, where we could hole up and regroup.

As the member of the Republican leadership responsible for communications, my immediate concern was how to reach members and keep them informed on what was going on. Before I had a chance to take action, however, the Capitol Police called to say the decision had been made to move the Republican and Democrat leadership to a secure location. Within minutes, the police were downstairs to whisk me away, first to police headquarters and then to a chopper waiting on the grounds of the Capitol's west front.

Along with ensuring the ability of the Congress to continue to function, communicating with members was a priority of the leadership gathering at the Capitol Police HQ. Before we lifted off, we instructed the communications staff to reroute all phone calls made to the Washington offices of House members back home to their district offices. This gave us a way to keep communications flowing in both directions.

I don't think I could ever adequately express my appreciation for the cool professionalism of those law enforcement officers at this most difficult of times. The police and FBI were tremendously efficient, and I soon found myself aboard a chopper with Democratic Minority Leader Dick Gephardt, Minority Whip David Bonior, and Republican Whip Tom Delay. Speaker Hastert, who is third in the line of presidential succession, had already been airlifted to the secure bunker, along with Majority Leader Dick Armey.

Our half-hour chopper flight gave all of us some time to meditate on the gravity of the situation and count our blessings. In times of great stress, you also find yourself thinking about the oddest things. Do I have enough life insurance? Did I hug each of the children before I left and tell them I love them? Did I tell Frankie "I love you"? The noise of the choppers kept us from talking with one another, but once we arrived we began to discuss the situation.

It wasn't long before the Senate leadership joined us in the bunker, too. We talked about contingency plans and how we might return to a semblance of normalcy, but for a few hours, there was little else we could do except wait for news. The CIA and the FBI came to the bunker to brief us on the latest developments, but details were sketchy. Vice President Cheney also called several times and kept us up to date on efforts to bring the president back to Washington. With a couple of planes in the air still unaccounted for, they were taking no chances with the president's safety, and no one disagreed with that decision.

Meanwhile, like most Americans, we watched the continuing news coverage on Fox and CNN, shocked by the pictures we were seeing of the terrible devastation. David Bonior, visibly shaken by the events, voiced what we were all feeling. "The sad thing is," the minority whip said, "this is real."

Yet we felt oddly disconnected from reality, estranged from our fellow members of Congress, worried about our nation, its president, and our own families. There we sat, miles from Washington in this sterile, vanilla gray room that seemed a throwback to another era, with its slightly outdated government-issue furniture and bare walls. As we talked quietly, my mind flashed on an earlier trip to another bunker. The previous November, the Gridiron Club, a social organization made up of members of the media, had invited me to speak to the group at the Greenbrier. It's a historic old mountain resort in West Virginia about four hours from Washington. During the Eisenhower administration, the government built a secret bunker 700 feet below the hotel's west wing to house members of Congress and their staff in case of a nuclear attack. No longer in use, Greek Island, as it was code-named, is open to the public; while I was there, I was given a quick tour.

It was eerie walking through the 25-ton bunker doors down the long hallway to the decontamination areas, realizing that had I been there as a member of Congress in another time, it would have meant the country was under attack. Now here I was 10 months later, in another bunker much like the one I had toured; but this time, our nation really *was* under attack.

Our time in the bunker lasted from around 12:30 until about 5 in the evening, when we all returned to the Capitol. Not long after our arrival, members of the House and Senate joined together on the steps of the Capitol to send a message to the world: This country is united, and the evil of terrorism will never defeat us. Republicans and Democrats, men and women of every color and religion, stood together, holding hands to sing the praises of our great nation. "God Bless America" has always been a favorite of mine, but that evening as the sun began to set on the shining dome that still proudly towered over the Washington horizon, the words of Irving Berlin's great patriotic tribute took on special meaning for each of us standing there on the steps.

God Bless America
Land that I love.
Stand beside her, and guide her
Through the night with a light from above.

The following evening I spoke to a joint prayer service in the Capitol Rotunda. That night, I didn't need a text. I was speaking from the heart when I told the members a story about one of my daughters' birthday parties. The kids were out playing with balloons when Kesha's balloon popped. All upset, she came running up to me and jumped in my lap.

"Fix my balloon, Daddy," she cried. "Fix it."

Well, I couldn't put that balloon together again. But what happened to us on September 11 could be repaired. Yes, our balloon had popped, but we could jump into God's lap and say, "Here, Father, fix it," and trust in him to help us put our world back together again. It was an emotional evening for all of us. I never doubted that God was with us at every moment during those long and terrible days right after the attacks.

With the nation's air service grounded, it wasn't until the following

Saturday that I was able to hitch a ride on a private plane heading for Arkansas and then on to Oklahoma. It seemed like years since I'd last seen Frankie and the kids. They were heavy on my mind, and when the plane touched down at Will Rogers Airport in Oklahoma City, I scrambled down the steps to the tarmac, got down on my knees, and kissed the ground. Then I thanked the good Lord for getting me there safely. Like Dorothy said in *The Wizard of Oz*, "There's no place like home," and after five of the worst days of my life, at last I was there.

The first thing I did was give everybody a big hug. I know the girls had been especially worried the day of the attacks, calling Frankie from school to make sure I was all right. Trey plays his fears a little closer to the vest, but once he saw me, he was fine. Frankie and I held each other a little longer and a little tighter. They were all a sight for sore eyes. The next night we had a family prayer session and talked all about the terrible events of the week. Just seeing my family and being with them for a couple of days gave me a tremendous boost and a needed sense of relief, but the following Tuesday evening, I got on another plane and headed back to a Washington that would never be the same again.

Believe me, I would just as soon have stayed put. No one likes uncertainty, and it's times like these which remind us that the burden of leadership is sometimes heavy. But you just trust that God's grace is sufficient and that God is still in control. Like all Americans, members of Congress had a job to do; and at that moment, the best thing we could do for our country was to get it done.

Congress went back to work. The president's leadership and resolve brought strength and comfort to a grieving, shell-shocked nation. And our brave military men and women stepped up to the plate, left their families behind, and put themselves in harm's way to preserve freedom and defeat evil. Their welfare was never very far from my mind and still isn't. They certainly were in our thoughts late last winter, when the president invited the members of the Republican leadership to join him at Camp David for an overnight stay.

We covered a lot of ground during our short visit, talking about the successes we'd achieved in Afghanistan at that point and the difficult challenges still to come. It was a serious meeting as the CIA briefed us on the latest intelligence. We also talked a good deal about the Republican domestic agenda and how we could work with one another and with our Democrat colleagues to get some things done for the American people.

I'd had meetings with the president before, and it is always an honor to spend time with the commander in chief. Usually, our paths cross at the White House or up on Capitol Hill, so I was excited to finally see the presidential hideaway I'd heard about for years. Camp David, named by President Eisenhower after his grandson, is a rustic retreat of simple cottages nestled in a thick forest located deep in the Catoctin Mountains of western Maryland; presidents have used the site as an escape from the pressures of Washington since FDR sat in the White House. In fact, Roosevelt and Churchill crafted strategy here during World War II, and it was at Camp David that President Carter conducted the successful negotiations that led to the Camp David Accords, the historic normalization agreement between Israel and Egypt.

Presidents have bowled here, shot skeet, snowmobiled, and pitched horseshoes. Little Caroline Kennedy even rode her pony, Macaroni, down the camp's wooded trails. It's a wonderfully peaceful place, and within minutes of arriving, you understand why most presidents develop a fondness for this old mountain hideaway.

We had a truly enjoyable dinner with the president, and then took up his invitation to join him for a movie. There's a small theater built right on the grounds. The president's film choice that night was an interesting one—*Black Hawk Down,* a tremendously sobering story of the personal sacrifice and extraordinary courage of the American military thrust into Somalia's bloody civil war. As I thought about his choice later, I suspected that there was method to the president's "madness" in choosing this particular film.

It tells the story of a group of Rangers and Delta Force commandos who found themselves surrounded by thousands of gun-toting rebels as they tried to capture a number of rebel leaders in one of the most dangerous neighborhoods of the capital city, Mogadishu. In the process, two Blackhawk helicopters were shot down and the pilots were lost. The Rangers' code, however, dictates that no man be left behind; and in the attempt to rescue the downed pilots and others caught in the fighting, a terrible and bloody battle occurred. American forces found themselves fighting block by block through Mogadishu's narrow streets and alley-ways as hundreds of well-armed snipers rained bullets down on them, roadblocks stalled their progress, and the presence of many civilians complicated an already deadly situation.

When our forces were finally rescued the following day, eighteen soldiers were dead and seventy-three wounded. Three days later, President Clinton announced the end of the U.S. mission in Somalia.

There were a number of reasons for the casualty rate, but one of those was a lack of armored vehicles that could have provided protection for the troops as they moved through the dangerous streets of that brutal city. We now know that those in charge of the Rangers' task force had requested Bradley fighting vehicles, Abrams tanks, and AC-130 gunships that might have saved many lives and prevented injuries. The decision to reject that request was made in Washington by then Secretary of Defense Les Aspin. Why? Because of "domestic political considerations."

The Clinton administration was fearful that arming our troops with this kind of state-of-the-art equipment might be interpreted as a significant escalation of hostilities, and that would have been too politically provocative for home front consumption.

President Clinton and Secretary Aspin had to live with that fatal miscalculation, and as the movie ended, I wondered if Mr. Clinton had seen it. If he hasn't, he should, in order to understand fully what his decision to put political considerations ahead of the safety of our soldiers cost in human terms.

This disaster had occurred before I came to Congress, but most of my fellow moviegoers had been in the House or Senate at the time. One of them was Senate Minority Whip Don Nickles.

After the movie, we rode together to our cabins and I asked him, "Don, you were around Washington when this happened, what's your reaction to this?"

He said, "You know, it just gives me the same ugly feeling I had back then, when I learned what Clinton and Aspin had done."

The next afternoon as Speaker Denny Hastert and I rode together to the landing field to grab a chopper back to Washington, he told me of the president's reaction. It seems President Bush had given Denny a lift back to the Speaker's cabin, playing chauffeur in one of the camp's golf carts. It's not a pickup, but not even the president always gets what he wants.

As they talked about the lessons of the movie, the president said the film doesn't tell us what to do, but it does show us what not to do. And he went on to say, and I'm paraphrasing, that he could guarantee his administration would not make the same mistake.

I found the film very disturbing, and for those of us charged with protecting the nation's security, I believe it was good for us to see the terrible realities of war. The human costs should never be far from our minds as we make funding and other decisions about this new war against terrorism. Thinking back, I don't think the president's choice of movies that night was a coincidence.

Although short, that time in Camp David was a welcome opportunity to recharge our batteries and renew our commitment to a just cause. Then it was back to Washington and the usual House business—budget debates, committee hearings, and meetings with folks from back home. But it wasn't business as usual. The attack on America had changed our world forever.

On September 11, after a year of vicious partisan feuding, the leadership of the Congress found themselves sitting together in that bunker not as Democrats and Republicans but as Americans. I hadn't seen many non-partisan moments in my years in Congress. This was something extraordinary. The partisan politics that had followed the presidential election disappeared in the smoke and rubble of September 11. But more than that, I believe our country underwent a fundamental shift as images and emotions of that day were forever seared on our national psyche.

Today, we see our role in the world differently. We see our responsibility for the lives of others in need through a new prism of pride and pain. And we understand far better what is really important in this world: family, faith, community, and country. In the months that have followed, the American people have made one thing clear. With the country at war, they want no return to the bitter partisan politics that have characterized Washington in recent years. And for a time, I do think both sides did their best to avoid partisan rancor.

But as I write this months later cracks in the commitment to a less divisive partisan atmosphere have begun to appear in Washington. People on both sides of the aisle and at either end of the philosophical spectrum are returning to the antagonistic behavior that unfortunately has been the pattern in Washington over the past decade.

On the left, many see problems from a more emotional point of view. They believe that their ideas and proposals are morally superior and that they are the anointed protectors of the common people against the hordes of elitist conservative infidels. In their world, all government spending is good; all tax cuts are bad and the only source of federal deficits. They never met a government program they didn't like, and help for disadvantaged Americans doesn't count unless it comes from the government. Hollywood is never held accountable for creating the increasingly valueless culture in which our children grow up, and group identity is more important than individual achievement. Their philosophy says if it feels good, do it. If you don't want to, don't. If it's a nuisance, abort it; if you can't handle it, drink it or drug it. If you don't like it, divorce it.

But that's only half the Washington problem. On the right, many see their mission as divinely inspired and themselves as saviors of America's traditional values from the evil forces of libertinism. In their world, all government social programs are bad. Compassion is weakness. Personal responsibility is the panacea for every problem. Their philosophy represents a black-and-white world in which there is only right or wrong; no shades of gray. Compromise is anathema, and strict adherence to conservative dogma is demanded.

In spite of what I think, or what the far left or far right think, we are not a right-wing or left-wing country. The American people exist somewhere in the middle, probably slightly right of center, refusing to embrace the ideologues of either side. They reject what they perceive as the often insensitive moralizing of conservatives just as they refuse to accept what they see as the moral relativism of extreme liberals. Yet the far right and the far left dominate news coverage and drive the political debate in Washington. I believe people are looking for something more from their elected leaders, something different.

They want solutions. They want a civil tone in Washington, less fighting and more progress. The American people have moved beyond their political leaders, and we need to catch up. We need to stop fighting yesterday's battles and quit blaming the other side for past wrongs and behavior, real or imagined, to justify our own actions. We need less stereotyping, fewer ideological confrontations, and the wisdom to understand that when a nation is as politically divided as this one has been in recent years, there are times when working with each other and building coalitions may be the only way to achieve progress.

Members of Congress in both parties have some changing to do to meet the new expectations of the American people and make this country all it can be. That doesn't mean Republicans or Democrats can't point out the differences between our parties' approaches to solving problems and argue forcefully for our causes. As a member of the Republican leadership, that's been part of my job description; and more to the point, it's what democracy is all about.

I am convinced that the principles of the Republican Party offer the best prescription for what ails us as a country. So, a strong, viable Republican Party is important to me as a conservative but also as a parent, an American of African descent, and a Christian. But if we want to grow the party for a better tomorrow, we need to ask ourselves some tough questions today.

Is our party listening hard enough to what America is saying? Do we really hear when people talk or are we just paying lip service with the expected response? Can we remain true to the traditions and fundamental principles upon which this party is based—freedom, opportunity, the rights and responsibility of the individual, and smaller government—and still be open-minded to differing points of view? Is it enough to be right? Do we try hard enough to understand how America is changing? Are we reaching out to people of every color, religion, ethnicity, and region to invite them to join our cause? Are we prepared to be a "governing" conservative party, or are we still mired in the "loyal opposition" mind-set?

When it comes to our future, I don't think the Democrat Party can get this country where it needs to go. I believe deeply that the values and ideas of the Republican Party will do far more to help far more people find the promised land if we are willing to lead. But leadership can be a tricky business. The Michigan State Police have a maxim: "The crux of leadership is that you must constantly stop to consider how your decisions will influence people." I don't think we do that nearly enough. Nor do we talk about our ideas in ways that connect with most Americans. We must understand that the politics of the next 20 years will be different from the politics of the last 20 years.

The Republican Party is terrific at determining how a program will impact the federal budget, but we're not nearly as good as the Democrats in explaining to people how our agenda will directly benefit them and their families, especially to minorities. We can assess the negative impact of drugs on the nation's economy in our sleep, but we rarely talk about the devastation they wreak on users and their families. When it comes to tax cuts, Republicans can quote budget numbers till the cows come home, but billion-dollar budget statistics don't mean much to most people. Democrats have an inborn talent for talking up social programs and the help they provide in very personal terms. To be as successful, Republicans must learn to translate tax cuts into human benefits, not billions of dollars, but more money for school clothes or a family vacation.

As Cool Hand Luke would put it, "We've got ourselves a failure to communicate." We must rethink our language and our strategies, and we must reapply ourselves. Our inability to make both an intellectual *and* emotional connection with the American people is a constant problem. A few years ago, the federal school lunch program had become a bureaucratic mess, wasting millions of precious dollars that should have been going to feed hungry kids. Republicans wanted to revamp the program and give state and local officials the flexibility to tailor the program to meet their needs. The Democrats, unable to break free of the status quo, screamed that we were trying to gut the program. Nothing could have been further from the truth, but when the cameras were rolling, they talked about starving children while we talked about "reducing waste" and increasing "local control." We should have pointed out that our reforms would mean more, not fewer, needy children would get a good, healthy meal every day. We didn't, and we lost that fight as Democrats and the media portrayed us as heartless bean counters who didn't give a whit about whether children went hungry or not. Unfair, but our own fault for not getting out our side of the story.

If we are to lead this country successfully in the years ahead, we've got to do a better job of communicating our compassion along with our competence. Earlier I asked the question, is being right enough? For many of my fellow conservatives, the answer is yes. They talk in technical terms and, not surprisingly, fail to connect with the people they want to help. And I do believe that most Republicans, like most Democrats, get involved in politics because they care about people.

Yet Republicans find themselves trying to lead without first establishing the credibility of their compassion with those they hope will follow. That's a recipe for eventual disaster. If people don't believe you care about them, they aren't going to believe in either your promises or your proposals.

Since September 11, I've spent a lot of time thinking about our country, my party, and its future. Over the past decade, Republicans have made tremendous progress, and we should be proud of our achievements. We've got a great story to tell. The challenge is to do a better job telling it.

But the world is changing quickly and in so many directions. If we are going to be the majority party of the future, we must not remain chained to the past. So after much soul-searching, I have come to the conclusion that the moment has come for a new conservative vision— one that preserves the basic values of our party but throws open the door to new ideas and new people. We must adopt a "New Conservative Strategy for a Better America," a three-part vision for the future.

First, we must renew our commitment to the nation's fundamental values and rebuild our core strength: the family. Second, each of us must do our part to make compassion and diversity central to American life. We have to reach out to conservative and moderate blacks, Hispanics, and other Democrats and Independents willing to join George W. Bush, J. C. Watts, and other Republicans with whom they share both goals and values. And finally, we must enthusiastically embrace new models and ideas to meet the challenges ahead. We can't govern in the 21st century the way conservatives did in the 1970s, '80s, and '90s by simply defining the left. We have seen the vision of the left, and it is no vision. But we must offer an alternative—a conservative vision of opportunity, quality education, and economic and retirement security for every American.

Through this vision, I am convinced that our party can make this a better place to live for every American; a better place to raise our children; to provide opportunity here at home and to advance the cause of humanity around the world.

FUNDAMENTAL VALUES AND THE FAMILY

So where do we start? With the basics—fundamental values and family. The wave of patriotism that swept America after September 11 is undeniable. Across the country, flags were flying from car antennas and front porches. As tragedy drew us together on that evil day, we learned in a graphic way the sometimes terrible cost of freedom, and many Americans, some for the first time, understood that the values and principles on which this country was founded are more than words in a civics textbook. Freedom, individual rights, democratic values, and self-determination drove the founding fathers to create a nation. The people of that new nation embraced a concept that said, "All men are created equal and endowed by their Creator with certain unalienable rights, that among them are life, liberty, and the pursuit of happiness." The validity of that dream has not changed, but we have. For many, the values that we grew up with—faith, family, hard work and personal responsibility—have become passé.

Special interest groups file lawsuits to keep a kindergartner from saying a prayer over her peanut butter and jelly sandwich in the school lunchroom. A teenage boy talks trash to a teacher and pays no price for his disrespect. Hollywood glorifies sex and violence to children while hiding behind the First Amendment and cloaking itself in political correctness. And every year, fewer and fewer people go to the polls as cynicism with our political system grows. Now, this tragic moment in history has given us the opportunity to reestablish our founding principles and values in both the conduct of our government and our daily lives. And we can start with the family.

President Reagan said, "There is no institution more vital to our nation's survival than the American family. Here the seeds of personal character are planted, the roots of public virtue are first nourished."

The Columbine massacre in the spring of 1999 was a brutal wake-up call for America. Something is terribly wrong when two "above-average students" from "good" families go to school one day—a day just like any other—and two hours later, twelve children and one teacher are dead. How did this great country filled with so many good people allow itself to reach a point where two young men become so warped that they lose any connection to moral reality? How could they have become the outcasts they believed themselves to be, and no one in authority, not even their parents, apparently, took their behavior seriously? How could this happen here?

Some say the answers are complex and confusing. I think it is easy to understand why we often seem lost as a society. What is hard is finding the right path to lead us out. Over the past four decades, we have traded values and virtues for a shallow, narcissistic, empty, consumer-driven culture, and the impact goes far beyond our children. We see it in the mother who drowns her two young sons because they have become inconvenient, or in the man who kills his estranged wife in anger and revenge while his children watch, or in the drive-by shooters who don't care who gets caught in the crossfire.

If there is one problem that keeps me awake at night, it is our drift into the dangerous gray void of cultural confusion that comes with the rejection of moral imperatives that are central to a healthy society. The American family has always provided the values framework upon which each generation of children build their lives, and it is the breakdown of the family that lies at the heart of our growing cultural problems.

There are many reasons why families in this country are under stress. Some are cultural, others are economic and political, but they all have catastrophic results: a welfare system that devastated family structure especially in the black community; divorce rates that have reached epidemic proportions. In fact, the percentage of families headed by a single parent increased 13 percent during the 1990s while the marriage rate went down 9 percent. At the same time, cohabitation without marriage went up 48 percent. The highly respected historian Lawrence Stone has said, "The scale of marital breakdown in the West since 1960 has no historical precedent."

The traditional extended family, with grandparents, aunts, uncles, and cousins all living in close proximity, is disappearing as work and careers lure people to new locales, often far from the family. The kind of family support group that was always there for me growing up in Eufaula—Mama, Daddy, Mittie, Uncle Wade, and Aunt Betty—is becoming an endangered species, and increasingly disconnected children are the tragic result.

One of the leading commentators on our culture, former education secretary William Bennett, put it this way: "I believe that the breakup of the American family is the most profound, consequential, and negative social trend of our time."

Don't get me wrong, there are plenty of strong, healthy families out there right now living by the rules, doing a good job raising their kids, and trying to help others less fortunate than themselves. And there are a lot of hardworking, loving, single moms doing right by their kids under less than ideal circumstances.

But there is an increasing number of children growing up with very little parental discipline and guidance, which is crucial for helping them develop self-control and for feeling that they are loved and cared for. No wonder so many look for love in all the wrong places. Those are often the only places left, and all of us are to blame.

It's not enough to take *our* kids to church, keep *our* kids away from corrupting movies, help *our* kids make the right choices if we look the other way when the secular institutions of American life—government, academia, business, and the arts—make the wrong ones.

Let me be blunt. The first step we can take toward restoring the American family is for each of us to recognize that we, as individual citizens, have all too often drifted with the cultural tide that has brought us to this unhappy place. We have not criticized our leaders when they veered from the values that made us strong for fear of being "judgmental." It's easier to commit an evil today than to criticize an evil.

We have not, for example, reached out to the many lonely and directionless kids in our communities, or sometimes even to our own kids. We have left the television on in order to entertain and baby-sit our children. We have fallen for the sound-bite mentality, which is a poor substitute for personal reflection and responsible action.

Make no mistake. We should not be surprised that some of our children are killing each other in our schools and on our streets. We should not be shocked that many of our children have no sense of purpose and feel hopeless and empty even in this land of opportunity. We should not be surprised that, in the absence of the faith of our fathers, our children have adopted bizarre creeds, beliefs, and practices. As a nation, we have refused to accept the fact that we are now reaping what almost every one of us has helped sow. And it is a very bitter harvest.

It is no coincidence that a society that undermines parental authority, marginalizes religion, and steeps its children in a violent and sexually obsessed popular culture produces children who are unruly, undisciplined, nihilistic, and in some cases infatuated with murder and quite prepared to act on these infatuations.

Clearly, too many children have been cut free from parental authority. When I was growing up, if Daddy reached for his belt, we knew all too well the meaning of the expression "hell to pay." Nowadays, a kid responds by threatening to call 911 and charge Dad with child abuse. Buddy would have howled at that notion. As the humorist Bob Orben says, "Life was a lot simpler when what we honored was father and mother rather than all major credit cards."

Parental rights and respect have been seriously eroded in so many ways by government, by movies and television, by liberal special-interest groups, and by parents themselves. Parental authority has also been undermined by easy divorce and public policies that have made fathers as rare as Steinway grand pianos in many inner-city homes.

Being called "Congressman" is a real honor, but it doesn't hold a candle to being called "Dad." I know that being a father has been the most challenging and rewarding job I have ever taken on, and the most important.

But fathers do more than bring home the bacon and mow the lawn. Father Theodore Hesburgh said, "The most important thing a father can do for his children is to love their mother." Dads are role models, teachers, and disciplinarians in ways quite different from those of moms. Today, in the majority of households, both Mom and Dad work, but each provides the children with distinct gifts. I've certainly seen it in our house. Frankie and I don't duplicate, but we reinforce each other's special contributions to our children's upbringing. It was the same with my parents.

I have always been inspired by General Douglas MacArthur's

thoughts about being a father. More than most, he knew the importance of keeping our country strong, but he also knew what should be important in every family man's life. He said, "By profession, I am a soldier and take pride in that fact. But I am prouder—infinitely prouder—to be a father. A soldier destroys in order to build; the father only builds, never destroys. The one has the potentiality of death; the other embodies creation and life. And while the hordes of death are mighty, the battalions of life are mightier still. It is my hope that my son, when I am gone, will remember me not from the battle but in the home repeating with him our simple daily prayer, 'Our Father, Who art in Heaven.'"

In the name of being nonjudgmental, the purveyors of political correctness now want us to believe that when it comes to raising children, there is no difference between the traditional mom-and-dad family and a single-parent family or a lesbian or gay home. And the national media, television, and movies contribute to that misguided notion by portraying every family situation as equal in terms of the welfare of a child. It's not. And there are plenty of studies out there to prove it. Sometimes single parenthood can't be avoided, and I know plenty of good, caring single moms like Frankie's mother who are bringing up wonderful children. But it's not the ideal, and most single parents will tell you, trying to raise kids alone is no picnic.

I was absolutely stunned to read recently that political correctness has gotten so out of hand that one Manhattan private school announced a new diversity-sensitive policy: it decided to ban Mother's Day and Father's Day. In the past for these special days, the children had made small presents in school for their parents. News reports said the decision to change its policy came after a gay man, who had adopted a child with his male partner, complained about Mother's Day.

The head of the elementary school explained the decision to reporters, "The reasoning was several-fold. One, [Mother's Day] didn't serve an educational need and, number two, families are changing. Some children were very uncomfortable."

As the *New York Post* put it so succinctly, "When did the biblical commandment—'Honor thy father and thy mother'—become a threat to children's emotional well-being?"

Unlike some in the conservative movement, I don't believe that bashing anybody accomplishes anything. I may disagree with the gay community's views and lifestyle, but bigotry and hate are not the answers to the differences that divide us but neither is a blind adherence to political correctness the way to restore values to our children or strengthen our families.

For many of us, our views and principles are the product of our upbringing. As I have said often in this book, my value system can be traced directly to a little town in eastern Oklahoma where my parents taught me right from wrong, personal responsibility, and the importance of family and faith.

The Watts family may not have had much by today's standards, but we were rich in the things that mattered. If most kids could grow up "poor" as I did, we'd all be a lot better off. Instead, too many children learn their values on television. Rich and poor alike are cheated of their childhoods growing up too fast with too little parental involvement. But perhaps most important, our children are forced to live in a relentlessly secular culture. They have been robbed of much of their religious heritage by the complete banning of religion from public life, especially in our schools. Don't think for a second that our kids haven't gotten the message. They can discuss almost anything in school except "Thou shalt not kill" and "Do unto others as you would have them do unto you."

Talking about the abiding truths of our Judeo/Christian faith makes too many adults very, very testy—sometimes to the point of legal action. And that sends a very clear signal to kids: Religion is either somehow bad or at least irrelevant to their lives. I deeply believe when these transcendent truths are driven from people's lives, negative results will soon follow. Contemporary history very much supports that assumption.

I am not saying that all our kids are bad. Far from it. For every troubled kid who gets his picture on the cover of a national magazine, there are millions more trying to make their way in this world in a decent and positive way. They are working in school and they are getting up in the morning wondering what they can do to make Mom and Dad proud of them. But we should all be concerned about the number of children growing up without a sense of right and wrong, anesthetized by a culture of materialism and moral relativism.

All this finally brings me to our entertainment industry. Let me be very clear. Hollywood creates many wonderful films. Some are nothing short of astounding, others are just fun. My kids and I watched *Beauty and the Beast, Home Alone* and *Remember the Titans* over and over. But this industry also bombards our children with epics of blood, sex, and moral ambiguity at best. Some are overt, such as Oliver Stone's *Natural Born Killers*. Others are much more subtle.

When it comes to Hollywood, our kids aren't getting the right message. Faith tells us that we were born to succeed, to achieve our goals, to be happy. Our children look around them and see killing, drugs, free sex, and corruption glamorized. Movies and television tell them sex without love or marriage is okay. We preach abstinence, and our teenagers see the president of the United States, who uses the Oval Office for sexual trysts, lionized on MTV. Record and video game companies tell them that violence is cool. And they see few people—in Washington or in the media or at home—willing to stand up and say this kind of behavior is wrong. Those are the messages our children are bombarded with every day.

Yet many in the entertainment industry, which makes billions of dollars a year from our children, will not accept any degree of responsibility for the outcome of their products. If movie producers and directors got the same treatment that corporate CEOs get when faulty products cause serious harm, maybe Hollywood might finally come to its senses. Instead, they cling to the totally discredited argument that children are not affected by what they read, see, and hear. This is obviously untrue, and any parent who has had to buy Pokemon cards, Beanie Babies, $80 running shoes, or carloads of Coke knows better.

Sometimes, Hollywood's artistic endeavors do change the world for the better. I truly believe, for example, that the entertainment industry played a positive role in improving race relations in our country. Remember *Guess Who's Coming to Dinner*?

But the time has come for Hollywood to admit that if it can inspire people to do good, it can also inspire people to do evil. And if it continues to create these valueless, degrading, and even dangerous films, it should pay a price. Am I suggesting censorship? No. Our freedoms should be inviolable. Some folks believe they can whittle away the First Amendment without threatening the Second. They are wrong. If Hollywood's First Amendment rights can be taken away, so can our Second Amendment rights. We should all be wary of toying with the Constitution.

What we can do, however, is expose, denounce, and boycott those who pump rot into our society. We must call them what they are. They are cultural polluters. They are playing a central role in the corrosion of our children's character, and they should no longer get a free ride. Yet the arbiters of political correctness look elsewhere when it comes to placing blame for increasingly violent children.

The Mall in Washington is a little like America's town square. On any given day, it is filled with crowds of people visiting museums or just enjoying a walk on the grassy carpet that leads from the Capitol down a broad expanse past the Washington Memorial and the Smithsonian to the White House. But on some days, the crowds are larger and noisier as protesters come to express grievances and demand change. A spring morning in May 2000 was one of those days as thousands of people, mostly women, marched down the Mall demanding that the Congress enact new, more restrictive gun laws in response to the Columbine killings the year before.

The Million Mom March attracted a bevy of Hollywood celebrities and musicians, led by talk show host Rosie O'Donnell. What I remember most, however, was the sign that one protester carried, which said: "Please protect us from guns." What she was really asking was for the government to protect us from ourselves.

This protester had so misread the problem of youth violence that she blamed inanimate objects for crimes that originate in the human heart. My view is vastly different, and was stated very well in an essay in *National Review* magazine. The writer of that essay made the point this way: "Blaming guns for horrors such as Columbine is no different than blaming the chains for slavery. The problem isn't the guns. The problem is what is in the hearts and minds of the very small minority that decides to kill."

Just like most Americans, I believe in sensible gun-safety laws, but not in an attempt to impose morality from without as a substitute for parental instruction or family structure. It is up to us to put this country back on the right track.

There are some ideas that we can pursue legislatively that can help us change course and move toward family renewal. We must work to return control of our schools to parents and communities, let parents make the right decisions about what's best for their children's education, provide for the inclusion of faith-based schools, and do more to help those kids who exhibit the early warning signs of violence and self-destruction. We must also provide for real tax relief, letting families keep more of their own money. It is wrong for both parents to have to work, spending less time with their children, purely to pay the family's tax bill.

I also think it's time politicians acknowledge that government simply cannot solve every problem within our society today. Government can be a positive force for change, but people are tired of made-for-television political rhetoric that promises an easy fix for everything from gun violence to poverty.

In this new, post–September 11 world, people don't want sugar-coated solutions that don't work. They want the truth, and the truth is, there is only so much government can or should do. In the end, rebuilding America's families can take place only in our own hearts and homes, where the breakdown begins. Nobody should expect Washington or Hollywood to say yes to changing their ways if we aren't willing to say no to our children more often at home.

We must never forget that we, as adults, have created the world in which our children are being raised. It is not our children who make the movies, video games, and television shows. It's not our children who have driven God from the schools. And our children are not responsible for the moral free-fall in our public and political institutions. We are. And only we can give them the world they deserve—a world free of hunger and poverty, where AIDS has been eradicated, and equality exists for every human being. A world of economic opportunity and compassion for those with differing views, cultures, and creeds. A world of moral clarity and conviction.

Our current cultural decay has taken decades to do its damage, and changes today may not be realized for years to come. But for the sake of our country, we must move our culture closer to one that prizes strong families and promotes positive cultural influences on our children.

Encouraging Compassion and
Celebrating Diversity

More than a century ago, Booker T. Washington said, "More and more we must learn to think not in terms of race or color or language or religion or political boundaries, but in terms of humanity. Above all races and political boundaries, there is humanity." Some truths never go out of style. Or shouldn't.

Encouraging compassion and celebrating diversity is the second part of our new conservative strategy to make America a better place. But both parties must acknowledge their responsibility for our country's continuing racial problems. Republicans have often been insensitive. Democrats and national black leaders have demagogued racial issues for political advantage. No one knows better than a black conservative the failings of both parties when it comes to racial healing. The glaring problems of race were never more evident than December 2000, after the most brutal presidential election in our nation's history.

On the morning of the historic Supreme Court decision that settled the presidential election, there was lively discussion under way on C-Span's *Washington Journal* program, America's electronic equivalent of the soapbox. It's a tremendous program that contributes mightily to political dialogue in this country and helps strengthen our democratic traditions. I've appeared on the show many times and always enjoy the give and take of the program's very interested and involved audience. It's not always fun to be on the receiving end of criticism, but for politicians, a stint in the hot seat every now and again is probably good medicine.

That morning, however, passions were running high. The country was badly divided along partisan lines, everyone believing their man had won and the other side was out to "steal" the election. As C-Span often does, it tried that day to make the discussion a constructive one by posing the question, "What should the new president, whoever he might be, do to unite the country?"

A Gore supporter called to say that he believed that Bush would ultimately be the president but argued that African Americans would never accept him as legitimate. Asked whether the involvement of Colin Powell and Condoleezza Rice would help bridge the gap between Bush and the African American community, the man strained for the politically correct words to express his real feelings. Finally, he said, "Colin Powell and Condi Rice aren't, uh, they don't reflect the African American community." I've heard this sentiment expressed on other radio and TV shows.

Could there be a sadder commentary on the state of black America today than this? That one of our nation's great heroes, a man of unquestioned integrity, courage, and compassion, who has dedicated his life to serving others, doesn't "reflect the African American community"? What can any of us do other than shake our heads in sadness that an articulate, extraordinarily talented, brilliant woman who has served both the highest echelons of our government and one of the world's leading institutions of higher learning doesn't "reflect the African American community"?

Why not? Haven't these individuals reached for the stars and through hard work, talent, and tenacity lit up the sky with their achievements and stature? Aren't these exactly the kind of role models children of every color need today?

Just about a month after President Bush took office, Julian Bond spoke to the NAACP's annual meeting. Rather than accept the hand that George Bush extended in friendship to all Americans in his inaugural address, instead of acknowledging the appointments of Colin Powell, Condoleezza Rice, and Secretary of Education Rod Paige, Bond chose to respond with harsh, hate-filled invective against the new president.

"They [the Bush administration] selected nominees from the Taliban wing of American politics, appeased the wretched appetites of the extreme right wing, and chose Cabinet officials whose devotion to the Confederacy is nearly canine in its uncritical affection," snarled Bond. Six months later at the NAACP's annual convention, Bond trotted out the same tirade and let loose again. The media, apparently not paying attention in January, treated his attack with the kind of coverage usually not accorded that kind of speech.

As a black man, I was embarrassed by Julian Bond and his brand of racial politics. The dogmatic, intolerant speech that was once the prerogative of white racists has now become the staple of some black leaders. I know. I have been the target more than once. So has Clarence

Thomas, Thomas Sowell, Bob Woodson, and Shelby Steele—in fact, any black who fails to follow the narrow philosophical paradigm of the liberal left. We are relegated to the status of ideological pariah.

But what does being black mean in America today? Does it mean I have to believe that O. J. Simpson was innocent? Do I have to believe in the principle of racial preferences? Can a black person oppose big government? Does it mean that I have to oppose the death penalty or parental school choice or the president's faith-based initiative? Do I have to be opposed to tax relief? What color is a conservative? Why can't a black man or woman espouse a more conservative viewpoint, take pride in independent thought, and still "reflect the African-American community"?

Chris Rock and I tried to sort that out one night in a lighthearted discussion about being a black Republican. I suppose Chris and I might have seemed the perfect odd couple—the conservative "faith-based" minister congressman and the rough-talking, irreverent comic. Nevertheless, mutual friends made the suggestion that we get together. They told me that although Chris can be wild and crazy, and he expresses his comedy in language that you won't hear in church, he really did have an appreciation for me. And in his own way, he really does have an appreciation for strong traditional values.

I enjoy good comedy, and Chris is rapidly becoming one of the greats. He's one of those comedians who take you back to your roots, to the characters and realities that we grew up with. In high school, my friends and I were crazy about Richard Pryor. I think you can trace a line, if maybe a shaky one, from Pryor to Eddie Murphy to Chris Rock.

For months, we'd tried to work out a mutually acceptable time for me to appear on his television show. Finally, I was scheduled to be in New York during one of his taping sessions, and the date was set.

I was expecting to get a good needling from Chris about being black and conservative, and he didn't disappoint me. But I had a few cards up my sleeve. I kicked off the conversation by telling him how much we had in common: growing up in homes where Daddy worked two and three jobs and we stretched every dollar.

"But what's that got to do with being a Republican?" he demanded. "Didn't you feel a little uncomfortable being the only black guy around?" he asked, referring to my speech at the 1996 Republican Convention, which he covered for Comedy Central. "There was no place to get hair grease or nothing."

I laughed out loud. "That's why I was looking for you. I was gonna borrow some."

I told him that I'd been the first before, and I had never chosen to be or not be something because of the color of my skin.

When he conceded that he'd be happy if one of his little brothers were to turn out like me, I was really touched, but that didn't get me off the hook. Chris started to good-naturedly push Republican buttons again on everything from welfare reform to what I did for fun.

So I talked to him about an issue near and dear to my heart that I believe will empower poor people—parental choice in education—and I reminded him that Republicans were leading that charge. He seemed to agree with me on that one, but I wasn't getting off that easy.

"Who do you hang out with?" he asked. "If George Clinton is coming to town, who do you call?"

"Who's George Clinton?" I should never have confessed my ignorance of the founding father of the Parliament Funkadelic musical axis. But, honestly, the last concert I can remember going to was when Frankie and I caught a Kool and the Gang show in Tulsa many years ago. I told Chris the awful, uncool truth: that I pretty much just hang out with my family.

"What did you do," he inquired, "when you were a sinner?"

"I watched *The Chris Rock Show*," I shot back with a big smile. Everybody got a laugh at Chris's expense. So did he.

I read recently that some people have suggested to Chris Rock that he enter politics. He replied that he wants to "hang out with Janet Jackson, not Jesse Jackson." Who can blame him? Politically, Chris is really a contradiction. In many ways, his views on responsibility and hard work are conservative like mine. On some social issues, we'd probably disagree, but he would face the same demands on his thinking from the national black leadership that every black politician experiences.

The truth is, these leaders fear and distrust any black who refuses to embrace their view of the world. Those who don't follow the group identity pay a heavy price as they are vilified; their blackness and their motives are questioned in the most personal language.

Simply put, many black leaders, today, have become the new arbiters of black orthodoxy. If we are African American, we are expected to obey their ideological commands without question. Should we stray too far from black orthodoxy, we are punished and isolated in a warped kind of ideological apartheid. I realize this is a harsh analogy, but when a fine man like Clarence Thomas is torn to shreds before the nation because he dares to think differently—to think outside the group identity—we are in danger of betraying what the civil rights movement stood for: the right of black Americans to control their own lives and destinies.

Sadly, this new black-on-black ideological "crime" is growing. As I was rushing back to Oklahoma for Thanksgiving in 1999, I was getting off a flight from Washington to St. Louis determined to make my connection to Oklahoma City. Congress had gone into recess a few days before. As soon as it did, I logged several thousand miles of campaigning for fellow Republicans in various cities, returning to Washington to tie up some loose ends. Just one more leg of jet travel would get me home ready for turkey with all the trimmings.

Waiting where my flight was disembarking was a black man a few years older than I wearing a uniform. He was an airline employee, so I stepped up to him and said, "Can you tell me where the gate is for Oklahoma City?"

He looked up, obviously surprised. "Oh, man," he said. "I didn't ever expect to see you here." Then he kind of chuckled and added, "Especially since I hate you."

"You *hate* me? Why in the world do you hate me?" I asked.

Realizing he'd put his foot in his mouth, he backtracked. "Oh, no, no, no. I'm sure you're a good guy."

I'm in airports all the time, and I hear lots of different comments. But this one really hit me between the eyes. I walked off to my gate wondering if I should have said something else to draw him out like, "Do you hate me because you think my being black means I should be a Democrat? Do you hate me because even though I'm black, I don't believe exactly what you believe? Do you hate me because I believe that I'm able to succeed without the government's help? Do you hate me because I refuse to live my life as a victim? And would you ever say the same thing to a white congressman?"

Double standards bother me very deeply. It's considered a given that Democrats, especially black Democrats, can't be racists. But isn't it racist to think that all blacks must have the same political philosophy? And wouldn't it show intelligence to learn about someone's different viewpoint instead of hating that person for being different? Isn't it possible that breaking away from dependence on government aid might bring people of all colors more self-determination, more self-satisfaction, a greater sense of managing their own destiny?

With these half-angry thoughts tumbling through my head, I reached my gate and went straight to the nearest phone booth. As I opened my book of contact numbers, a young black man came striding right up to me. He was about 20, a Howard University student heading home for his first family Thanksgiving in three years. It was also his grandfather's sixty-fifth birthday, and he was excited about having the opportunity to share in that celebration as well.

"Congressman," he said as he shook my hand, "I really admire you for being your own man. I'm an economics and political science major, and I appreciate what you're doing."

What a difference a few minutes can make.

Socially and politically for the past 40 years, the acceptance of black victimhood by both the black and white communities has produced unintended results. As a society, we've been mining a big lode located right at the point where white guilt and black anger intersect. And there's plenty of material there to be excavated. For those who cling to the identity of being a victim, however well-justified that stance might be, it puts a ceiling on personal development. And for the guilt-driven, it leads to patronizing, ultimately dismissive solutions, government-dependent solutions that limit understanding, opportunity, and real equity.

After my State of the Union response, the *New York Times* noted my non-traditional Republican stands on affirmative action and civil rights with an editorial titled "The Trouble with J. C. Watts—He Thinks for Himself." I appreciated the compliment, but my independence goes both ways. Yet my differences on issues with traditional black leaders rarely earn such praise from the media.

Many of these leaders come from roots very much like my own.

Many of them once stood alongside me on the sanctity of life and of marriage and in the belief that responsibility and hard work create better lives than handouts or entitlements. Perhaps, it is they who have forgotten where they came from. Many have traded results for a seat at the table. Many have made government their god. Many have deserted the crucial lesson of self-sufficiency that came out of their upbringing. In doing so, they've abandoned the most fundamental human needs of the people they purport to serve: the need for self-determination and self-respect.

Privately, many of these leaders would acknowledge that in the end welfare hurt people more than it helped and would agree that affirmative action has helped only certain blacks. But publicly and politically, these facts are not convenient, so they dismiss them.

The character-building and opinion-shaping experiences of my life have helped me see beyond political orthodoxy and beyond the constraints of group identity. My formative experiences led me to embrace a set of principles that can totally reorder the social and political landscape of America. Yet I am expected to walk away from my beliefs because I'm black. And my blackness is questioned. I would pose this question: Who has earned the right to speak for the African American experience, a black man raised on the poor side of the tracks in Eufaula, Oklahoma, where poverty was a way of life and whose first schools were segregated? Or a man raised in a world of affluence and power who attended one of the nation's most elite private schools?

I am not suggesting that Jesse Jackson Jr. or other blacks raised in privilege don't have a perfect right to their liberal views. They do, and I respect them. My complaint lies in the unwillingness of most black leaders and the media to accord me the same legitimacy to hold conservative views even though I was raised in the poverty others have only studied.

I raise this issue of ideological apartheid because it is symptomatic of the racial problem that is beginning to erode the racial progress we have worked so hard and long to achieve. As black leaders demand allegiance to group identity, they fail to see the harm they are doing the movement and the country. This group identity can sometimes limit us more than protect us. It can blind us to valuable viewpoints, options, and opportunities. To maintain political power, the group identity must remain ideologically pure. So we heard terrible hate-filled radio ads in the 1998 election aimed at inflaming racial fears in black communities. We saw both President Clinton and Vice President Gore use race not as a uniting force within our country but to cynically divide the races for political gain in a way we haven't seen in decades. In the 2000 election, the infamous Byrd spot tying George Bush to that horrific Texas murder of a black man took racial fear-mongering to a new low. We cannot hope to bring this country together if black leaders continue to embrace the politics of victimology, the politics of fear, and the politics of ideological apartheid.

Nor can we become one America if some of my conservative friends refuse to see that racism remains a stain on this nation's soul, and poverty, regardless of its root cause, is reason for concern. It's ironic that the party of Lincoln has often been absent on issues of civil rights and equal opportunity, failing to understand the historical sensitivities of the African American community.

The political aftermath of the Florida recount is a perfect example of a lost opportunity—the result of both the irresponsibility of national black leaders and the insensitivity of some Republicans to black concerns. While it is true that no evidence has been found of any organized conspiracy to disenfranchise black voters in Florida, we've seen an emotional backlash against President Bush and Republicans by many African Americans. In many black communities, the few isolated instances of black voter complaints have taken on near mythological proportions, spurred by the heated rhetoric of national black leaders.

Republicans argue correctly that almost all voting irregularities took place in counties where the Democrat Party controlled the election process. But what African Americans hear is, "We were right. Case closed." I certainly agree that questioning the legitimacy of an election —to further divide our nation—is wrong. But rather than trying to understand this terribly negative response and look for ways to counter what is clearly a campaign of misinformation, many Republicans argue that it is pointless to even try to reach out to African Americans, and that attitude is also wrong.

If we are going to convince African Americans that what happened in Florida was not a denial of civil rights (and it wasn't), then in the future we must stand with them when *real* cases of discrimination are found. If we are going to connect with people, we must not only provide solutions to their problems, we must be with them in the trenches where minority children still try to learn in second-rate schools and single moms struggle to pay the rent, put food on the table, and cover child-care costs. Furthermore, we've often sounded more concerned about principles and programs than the people they are designed to serve.

As Republicans, we can't fulfill our responsibility to promote diversity by demanding a robotlike allegiance to every conservative view. Making opposition to affirmative action a test of ideological loyalty is as wrong as making its support the prerequisite of political authenticity in the black community.

When I spoke to the nation in my response to the State of the Union address, I used the last few moments to talk about the racial divide that still exists in this country. I told the audience I was just old enough to remember the Jim Crow brand of discrimination. I've seen the issue of race hurt human beings, and our entire nation. Racial healing is an issue we can't afford to ignore if we are to remain "one nation under God, indivisible, with liberty and justice for all."

Too often, when national black leaders talk about racial healing, however, they rely on the old assumption that government can repair the racial divide. That's understandable. It took the passage of some great and good civil rights laws to bring this nation out of the pervasive racism that kept me sitting in the balcony at the Eufaula Theater well into the late 1960s. I know I have personally benefited from those laws, and I feel a huge debt to people like Dr. King and Congressman John Lewis, my father and my uncle Wade, who risked their lives so that I might be part of the first generation of black children given a real opportunity to achieve our dreams.

But I believe we're at a point now where we have to ask ourselves some soul-deep questions: If legislation and government programs are the answer to the racial divide in our nation, then why, in our time, does the division seem to be growing? Why is the healing we long for still so far from reality? Why does it seem that the more laws we pass, the less love we have?

The fact is, our problems can't be solved by legislation alone. This has been a long, difficult journey for our country, and we must now accept a great truth. Government can't ease all pain. We must deal with the human heart. We must individually accept our share of responsibility. It was once said, "Sorrow looks back, worry looks around, but faith looks ahead." As we stand at the beginning of a new age, we must decide whether we will continue to be held captive by the past or move on to a new era of racial co-existence.

This country must be a place where we all—red, yellow, brown, black and white—in some way feel a part of the American dream. I know many African Americans feel that racism is still a strong, negative force in this country. Many whites don't. I believe the truth lies somewhere in between. I don't think race relations are as bad as Al Sharpton and Jesse Jackson or others would have us believe—other than in the political trenches. I have preached in many churches where the choir looks like the United Nations. I have been in plenty of communities where people of all colors seem to be participating in the American process. It would be naive to think, however, that bigotry has become extinct in America. It hasn't, but we won't lessen the racism that still exists by further dividing this nation into racial groups or trying to stir racial animosities for a few votes on either side.

Not long ago, I stopped in at a 7-Eleven in Oklahoma City for a little gas when a black man in his late twenties came up to me.

"Aren't you Senator Watts?" he asked.

I laughed at the promotion and told him who I really was. This young man had stopped me because he was troubled and wanted me to enlighten him.

"You speak for so many of us," he told me. "A lot of my friends."

Then with a troubled look, the man said he feared that most blacks would shun him if he spoke up and acknowledged his conservative views.

"I'm a Christian, and I understand where you're coming from," he said, "but I am ostracized by so many for my views, even in church, even by my pastor."

"How do you deal with that?" he wanted to know. "What can we do?"

There is only one thing we can do. Both sides—black and white, Republicans and Democrats—must reject old prejudices if we are to make progress toward one America. I am a man of faith and hope who believes in the goodness of people at heart and their ability to reach out to one another without rancor. I know it is possible even between the most unlikely people in the most unusual places.

Last summer, I made my third trip to Africa. I led a delegation of congressmen, black and white, to Senegal, Ghana and Nigeria in west Africa. For me, it was a journey to my roots but it was also a mission to better understand our country's relationships with African nations and the terrible problems of poverty and disease facing many of these countries. In Kumasi, Ghana, we participated in a tradition that dates back thousands of years: Congressman Wes Watkins and I met with the king of the Ashanti tribe and his court. His Majesty Otumfuo Osei Tutu II is a big man, educated in Britain, and only 48 years old, young enough to embrace capitalism but old enough to understand and respect the traditional ways of his nation.

The audience was gathered in an open field in Kumasi, the sacred ground for royal meetings, as five men each held a 15-foot-wide umbrella made of African kente cloth. Together, the umbrellas made a canopy of color that was breathtaking to see.

At first, our meeting was a highly choreographed exercise of bows and conversation conducted through interpreters. The king generally doesn't speak directly to "commoners." Using the interpreter, he asked me to speak to his court and a crowd of villagers gathered outside. Just before I approached the microphone, it dawned on me that this wasn't any old speech. As I squinted into the hot sun beating down on the hundreds of black men and women standing before me, I realized I was standing on sacred land, the land where my ancestors had once lived as free and independent people before slavery changed their world forever. It was an overwhelming moment. And I told the audience, with whom I shared this ancestry, that the phrase "There's no place like home" now had new meaning for me.

The crowd erupted, and King Tutu did something quite extraordinary. He ordered one of his staff to hold an umbrella over my head to protect me from the burning African sun, which, I was told, meant that by Ghanaian tradition I was no longer a commoner. I suspect that's the closest I'm ever going to get to royalty.

Then something even stranger happened. A news photographer from the *Washington Times*, covering the event from the crowd, requested a picture of King Tutu and me. One member of the king's entourage approached him with the request. I was ready to head over to where he stood when he announced through his interpreter, "The congressman need not come to me. I will go to him." People all around me were whispering and shaking their heads in surprise as the king strode over and greeted me. I thought how different his world is from mine, yet our shared ancestry created a tie between us that overcame tradition and time. We were just two black men, proud of our people's history. It was a special moment for me, a connection I never expected to a world I'd never known.

But it was another stop on this trip that reinforced my belief in the idea that, regardless of our differences, each of us has the ability to reach out to others and build bonds of love and respect that can defeat evil and prejudice. Gorée Island, all 45 acres of it, is a rocky piece of land off the coast of Senegal. Beginning in the 1500s, 20 million Africans passed through this little island on their way to slavery. Dutch and other traders imprisoned men, women, and children in a bleak fortress that still sits high on a hill overlooking the sea. Today, a small market with people selling traditional African clothing, jewelry, baskets, and other trinkets exists where slaves in chains were once marched onto waiting ships.

As our group was taken through the holding areas of the prison, we saw 8-foot-square cells where as many as thirty men were kept before boarding ships bound for the New World. We saw where the children were housed after they were separated from their mothers, who were in cells of their own close enough to hear them cry.

Walking through this stone prison, the reality of the fear, anger, and despair these people must have felt swept over me. The stories of slavery that I had studied in school were no longer just mere historical accounts for me. They had become the painful chronicles of my own family.

As we moved from room to room, many in our group felt an almost palpable sense of dread. At the end of the tour, we walked down a dark hallway with a door at the far end. From a distance, however, all we could see was a blinding blue rectangle as the sun burst through the door into the dark hall. Through the doorway was a stone path that led down to the beach, which I was told the slaves called the "door of no return" or the "farewell to Africa door." Those who passed through it never saw their homeland again. Just as my ancestors had once done, I stepped through the door and paused to look down at the ocean that had carried so many people away to slavery. Then I realized that one of my fellow congressmen on the trip, Peter Hoekstra, had come up next to me. Pete has been one of my closest friends and allies in the Congress since I came to Washington and is a dedicated public servant of courage and integrity.

Although Pete was actually born in Holland, today he is the congressman from Michigan's Second District; it was Pete's heritage that suddenly crossed my mind as I grasped the irony of the two of us standing together at this particular spot. Here we were, a descendant of slaves side by side with a descendant of those who had enslaved them. Suddenly, I felt tears sting my eyes. It was a truly moving experience, as if the Lord had brought us together at the moment to erase any doubts I might have had in the ability of our nation and all its people to bridge the racial divide that still separates us.

My uncle Wade knew about bridging that divide better than anyone. About a month after Wade died in December 1998, I spoke to the National Press Club, which is one of Washington's major speech venues. I was still feeling a sense of loss, and I suppose that's why I decided to share some thoughts about my uncle with the audience.

"He was a gentleman," I told the reporters near the end of my speech, "who stepped in when I made a bad choice." But Uncle Wade's life was dedicated to the proposition that we could all live together in peace if our hearts were right. His years heading the NAACP in Oklahoma brought him death threats, but nothing ever stopped him from doing what he believed was right.

Then I told them the story of Uncle Wade and the klansman. It seems that Uncle Wade once accepted an invitation to debate the grand wizard of the Oklahoma Klan on local radio. During the argument, the man called my uncle every derogatory name in the book. Wade answered with spirit but never with personal attack. As strange as it seems, somehow, Wade's humanity that night touched something deeply spiritual in his adversary, and in time, the klansman became a changed man. Not only did he leave the KKK and embrace Christianity, he and my uncle became dear friends.

Just the month before my Press Club appearance, I sat in a pew at Uncle Wade's funeral as a long line of friends and admirers honored his passing. One of the many who spoke at his funeral was that former Klan member. It was truly amazing to see how God had worked in this man's life. I watched as he admitted, with tears in his eyes, that Uncle Wade was the first black man he had ever shaken hands with, testifying to how my uncle's great humanity had impacted his life.

If my uncle and the grand wizard of the KKK could find healing and unity, I believe all of us—black and white, red, yellow, and brown—can live together in harmony and love. We still have a long way to go in erasing the scars of racism and bigotry, but we cannot be intimidated by the distance between where we are and where we need to be as neighbors and as a nation.

If there is any good that came of the terrible tragedy of September 11, it is that we saw ourselves for the first time in a long while not in colors, or groups, or classes, but as one country. The evil men of Al-Qaeda murdered indiscriminately. People of every color and many religions—Christians, Jews, and Muslims—became targets of their hate. That shared experience has changed America in many fundamental ways, not the least of which is a new sense that we are, in fact, one people proud of our individual heritages, but Americans first. The group identities that have widened our differences over past decades suddenly seemed dated and inappropriate after September 11. But as we move on with our lives, I fear, many on both sides—black and white—will return to the anger, resentment, and racism of the past. We must not let the terrorists claim another victory.

Racism is a sin of the heart. We can treat some symptoms of racism. We can punish those who violate the rights of others, but we can't totally eliminate bigotry by putting a new law on the books. We can force people to go to the same school and eat in the same cafeteria, but we can't force them to love one another. Yet we should love each other. Love is the closest we can come to expressing the divine, the finest thing we can do with our time together on this planet. And with faith, it is the most precious legacy we can leave our children.

Harry Pearce, the former vice chairman of General Motors and now chairman of Hughes Electronics, has been honored many times for his leadership in protecting and promoting the civil rights of all Americans. He tells a wonderful story about a lesson he learned from his father when he was a young boy of 7 or 8. His father was the head of a small law firm in Bismarck, North Dakota.

As Mr. Pearce puts it, "We didn't have affirmative action in Bismarck back then. In fact, we didn't have any black people, period." So young Harry was surprised one day when he saw an elderly African American gentleman walk into the door of his father's firm, past the secretaries and right on into his father's office.

Later Harry asked his dad who the man was. "He is a good friend," his father said, "a Baptist minister who comes to see me every year. We sit in my office and talk about life and what we believe in."

Harry Pearce says about that lesson learned, "My father taught me something about respecting diversity that day. I learned, as a young boy, that the color of a person's skin didn't matter to my father and shouldn't matter to me."

If only all of us could learn that lesson as children the world would be a better place. We don't all have that opportunity, but we can listen to the wise words of Martin Luther King, Jr.

"Our cultural patterns are an amalgam of black and white. Our destinies are tied together. There is no separate black path to power and fulfillment that does not have to intersect with white roots. Somewhere along the way, the two must join together, black and white together, we shall overcome, and I still believe it."

EMBRACING NEW MODELS

My last strategy to make America a better place—embracing new models to solve old problems—probably sounds simplistic to some, but you'd be surprised how difficult it can be to get members of Congress to throw off years of "groupthink" and come at issues such as education and poverty, trade and foreign aid, health care and retirement in completely different ways. Sometimes I think it would have been easier to get Buddy to part with a dollar than to get some of my colleagues, especially my friends on the other side of the aisle, to listen to a new idea.

There's a great puzzle that illustrates the mind-set I'm talking about. Look at the sticks below. The challenge is to remove three and leave four.

Most of us search for the solution and fail because we look at the sticks as a group of objects. But the solution lies in seeing the puzzle from a completely different perspective—as a graphic. When we do that, the answer is surprisingly simple. It's the Roman numeral 4.

I'm not suggesting that the answers to such difficult problems as poverty and health care and education are simple. Far from it, but I believe the answers will only be found, like the puzzle, through a different perspective. We must be willing at least to consider solutions that don't follow the traditional methods of problem solving because we know most of those approaches haven't worked in the past.

The Community Renewal Act is an example of the kind of new model thinking I'm talking about, an innovative approach to one of the country's most intransigent problems: inner-city poverty.

Welfare reform is beginning to impact our inner cities positively, but it is still true to say that America's urban areas have declined to a point that only a generation ago would have been unthinkable. Large numbers of people in our cities have always been poor, but today we are seeing a devastating disintegration of private society, beginning with families and neighborhoods, the institutions that nurture and discipline children, reinforce private virtue, and give order and meaning to daily life.

From violent crime to drugs to poverty, the quality of life in almost every quantifiable measure has worsened for low-income families over the past four decades. Defeating this assault on our cities and poor communities must be a national priority if this nation is to remain true to its promise of equality and hope. Our country has faced crises in the past. The year was 1968, and Americans were building new homes, buying new products, creating new businesses, and generally enjoying an unprecedented prosperity. The national economic atmosphere was heady and exuberant.

But on May 21 of that year, millions of Americans sitting in front of their television sets were shocked by a report from the respected newsman Charles Kuralt entitled "Hunger in America." That program exposed an unseen hunger and malnutrition that marked the lives of millions of Americans. Their fellow citizens were shocked into action, and ending hunger in America became a critical national goal. One editorial writer, commenting at the time on the documentary, noted, "The contrast of a rich country harboring pockets of the most primitive want was its own editorial on the social contradiction of an affluent nation."

Now it is more than 30 years later, and there is a new social contradiction—a new unseen hunger in the midst of prosperity. It is a hunger for opportunity, and it comes from our nation's poorest communities. It comes from the aging, struggling communities that most Americans have never seen; neighborhoods that have been bypassed by the national economic success story. These are the communities that cannot attract businesses and industry, to bring jobs and opportunities that lead to the American dream.

These are the neighborhoods where vacant properties become home to crack users who destroy the sense of safety and security that a community needs to grow and prosper. They are neighborhoods where potential business sites are neglected because of the high costs and even higher potential liability of environmental cleanup. They are city neighborhoods where a long public transit ride is the only way to get to the good jobs in prosperous suburbs. And in these economic wastelands, the people who live there don't have the opportunity to save for a first home, a new business, or a child's education.

I have spent the last eight years looking for new models that could address the intransigent problems of these impoverished communities. The result of my journey is the American Community Renewal Act, which was signed into law on December 21, 2000, after a long battle for passage. This legislation reflects American ideals by empowering individuals to renew the institutions of private society, which hold communities together through time-honored values of faith, family, work, community, and neighborhood.

The lessons I learned when I grew my entrepreneurial wings back in the 1980s were invaluable to me in crafting this bill. I knew what a small businessperson faces trying to get a business off the ground. It's hard for entrepreneurs everywhere these days, but add poverty, crime, and a lack of economic development to the mix, and you've got a recipe for failure.

The best antidote for failure and the greatest resource we have is our people—their hopes and dreams for themselves and their children, their ideas and ambitions, and most important, their goodness. And the poor people who live in these neighborhoods are no different.

By providing opportunities for low-income families, by encouraging businesses to create new jobs in low-income communities, by rewarding companies that clean up contaminated sites, and by forcing the federal government to localize ownership of abandoned housing, the American Community Renewal Act can help Americans lift the weight of hopelessness and rise up and achieve their dreams.

Despite the strongest economic growth in this nation's history over the past two decades, too many people living in our poorest communities remain trapped in an unbroken cycle of poverty. By bringing new businesses into these communities, the American Community Renewal Act aims not only to expand job opportunities, but also to provide opportunities for local residents to spend their hard-earned dollars right in their own neighborhoods. Once commercial centers can establish a foothold, seemingly destitute inner cities are often gold mines of consumer desire and potential retail activity. Magic Johnson, the great Los Angeles Lakers star, certainly knows that. He made believers out of a lot of people when he opened a chain of movie theaters in urban areas around the country and turned what many might call a risky business opportunity into a blockbuster investment. Nearly 2 million people go through his theaters' doors in Los Angeles alone every year. Of course, not everyone starts out with Magic Johnson's million-dollar name or bankroll, but I believe the opportunities to succeed are out there for people who are willing to work hard and who believe in themselves.

My daddy saved every dime he could, and when you're as poor as we were, his financial security later in life was a real testimonial to his tenacity and common sense. While he was never a rich man, my daddy understood that savings and investment were his ticket out of poverty. We need to help low-income Americans learn the same lesson, to allow them the opportunity to save and invest in their communities and in their children just as my father did.

I know from my own family history that owning property is a great foundation from which families can climb their ladder of success. My grandfather, Charlie Watts, born to survivors of the Civil War, paved the way for generations of Wattses to come with the cotton-growing acres he worked so hard to buy. My father, who purchased his first house as a teenager, learned the lesson of property from old Charlie, and his children learned the same from him.

When people own their own home or business, they have a personal stake in their neighborhood and community that makes all the difference—in how they see themselves and their responsibility to that community. They have abundant reason to care, to hope for the best, and to work for it. It's not just the investment of their earnings; it's the investment of their pride and their hope to leave something for the next generation.

The American Community Renewal Act embodies these ideas. Through this legislation, forty "renewal communities" will be established with targeted pro-growth tax benefits, regulatory relief, brown fields cleanup, and home ownership opportunities. Over the next few years, Congress will have to assess the program's effectiveness and learn from both the successes and failures in order to create new models for future efforts to extend prosperity to all of America's communities.

Passing the Community Renewal Act was no walk in the park. At times, it reminded me of that terrible Texas-Oklahoma game when I just couldn't get the ball down the field. It took 5 years, a lot of educating and cajoling, and a little compromise to finally get the job done. And why did this creative commonsense approach run into so many roadblocks? On the left, Democrats had a hard time accepting any antipoverty program that didn't have the federal government and a lot of bureaucrats running the show. They clung to old programs that have created their own constituencies—liberal special-interest groups, key players in the Democrats' political coalition that depend on these programs for funds and jobs.

On the right I had less trouble, but some of my colleagues rejected the idea of any new approach to fighting poverty that involved the government. Despite the emphasis on partnerships and individual empowerment in the legislation, they oppose most federal programs out of hand.

Overcoming the opposition was an instructive experience. I learned that too many members of Congress, particularly those on the left, have no interest in changing their approaches. Their solution is to just refund and tinker with programs that have consistently failed the country and the people they were meant to help. I always chuckle when I hear a liberal TV pundit expound on the unwillingness of conservatives to change. It's true that our fundamental principles don't change. But any truly objective observer of the Congress over the past decade would have to acknowledge that it is the Republican coalition that has been open to new ideas and innovative approaches. It is our party that has rejected the status quo. Conservatives, not liberals, have embraced the values of creativity, entrepreneurship, and private-sector participation in tackling tough issues because the old approaches simply weren't working.

The American people agreed. They knew kids weren't learning despite the billions of dollars spent by the federal government every year. They saw deficit spending had become a way of life, and so had welfare. What changed? The Congress. After the 1994 election, Republicans had the opportunity to finally throw off 40 years of Democrat control and four decades of business as usual. The result was a balanced budget, welfare reform, and now education reform, to name just three.

We haven't done everything right. After 40 years in the wilderness, we've had some learning to do especially in how we communicate why our agenda is better for people and for America. But we've shaken up the old guard that has ruled Washington for decades. Now we need to press harder to ensure that the momentum of the past 8 years continues.

We need new ways of restructuring our institutions so that the link between effort and reward becomes central to their mission again. New models based on more competition and less bureaucracy. Models that leverage federal dollars with state moneys and private-sector contributions and involvement.

It's time for corporate America and our government to work better together. Of course, we must hold businesses accountable for their actions. If a company breaks the law or violates the trust placed in it by employees and stockholders, we ought to throw the book at them. But at the same time, we should leverage the talent, connections, technology, entrepreneurship, and financial support business can bring to the table.

And we also need to assure the inclusion of some of our best sources of partners and problem-solvers—members of America's faith-based community. President George W. Bush said, "Private and charitable groups, including religious ones, should have the fullest opportunity permitted by law to compete on a level playing field, so long as they achieve valid public purposes, like curbing crime, conquering addiction, strengthening families, and overcoming poverty." I couldn't agree more.

Faith-based and community organizations have been quietly feeding the hungry, clothing the poor, housing the homeless, and fighting the evils of drug and alcohol addiction for many years. We all know that, and we know that often their results have been far superior to their mostly governmental counterparts. In much the same way that many faith-based schools educate children better on less money, faith-based organizations always seem to manage to do more with less.

We ought to be promoting the good work they do and empowering them with resources to reach out to those who need our help. But in the past, when it came to faith-based organizations, government regulations prohibited many from competing for the federal funds that could help them reach even more people in need.

Many of us who have been involved with these organizations and frustrated by the lack of progress achieved by governmental solutions, think it is time to "rally the armies of compassion" to America's side. Separation of church and state is a bedrock principle of our Constitution and must always be respected and protected. I believe that, and so do the majority of my fellow Christians. But it is possible to build a partnership between faith-based organizations and our government that lets the light of faith shine in the lives of the less fortunate without violating this important doctrine. We can and should let faith-based organizations compete for federal and state funds to help underwrite social programs that work.

Unfortunately, some on both the left and right oppose any efforts that would give faith-based organizations equal standing with their secular nonprofit counterparts or with government programs. Democrats don't see that these people, working in the trenches and suffering with those who suffer, understand compassion. Instead, they believe compassion is better dispensed from a safe distance by a faceless bureaucrat sitting in an air-conditioned office in Washington, D.C.

Some conservatives in the Christian community have also risen in opposition, arguing that making the federal government a partner in their efforts is like inviting the fox into the chicken coop. It's only a matter of time before government regulations and meddling would destroy the mission and the movement. With all due respect, they're wrong, and my friends in the Christian community need to understand that they are indulging in the same narrow thinking as their counterparts on the far left.

Every battle I've fought, every time I've run into another closed mind, I am more convinced that people must pressure their elected leaders to start thinking about issues in new ways. Asking ourselves fundamental questions about how we're going to approach the challenges ahead ought to be a part of our job description.

Let me give you some examples.

Social Security—Is Social Security reform about preserving the current system that is going broke, the status quo, or about ensuring a good quality of life for today's seniors and tomorrow's?

The environment—Is it about empowering trial lawyers and placating special interests or cleaning up brown fields that have turned inner-city neighborhoods into economic wastelands?

Medicare—Is it about perpetuating a bureaucracy or guaranteeing affordable prescription drugs for seniors?

Abortion—Is it all about *Roe v. Wade* or creating an environment that makes abortion unnecessary?

Agriculture—Are we concerned with agricultural economics or feeding the world?

Trade—Is it about opening markets or creating jobs at home and advancing humanity abroad?

Democrats and the liberal left are mired in the past, stuck trying to address the first half of those questions, while Republicans are trying to move beyond the status quo to consider these big issues from fresh perspectives. The outcome of this struggle will likely determine whether this country continues to take the comfortable, beaten path that is getting us nowhere or opts for the "road less traveled," which requires us to take some risks but might just get us where we need to go.

There is one more roadblock, however, that we must overcome on our journey to find new models. While many politicians in Washington remain wedded to old ways, to outmoded ideas and ineffective processes, many more cling to partisanship and ideology above all. They duel for political advantage while a news-hungry media, concerned with ratings and revenues, focuses its attention on the extremes rather than on the middle ground, where solutions will be found.

I am not suggesting that either side abandon its philosophical beliefs. Members of both parties should offer their best ideas based on their fundamental principles but in a constructive fashion. Nearly 40 years ago, President Eisenhower complained that he didn't like people who "go to the gutter on either the right or the left and hurl rocks at those in the center." After 44 years on this earth and after nearly 8 years in Washington, I've come to the conclusion that cooperation, rather than confrontation, is more likely to get this country where it needs to go.

Political reality dictates that with a five-seat majority, Republicans are likely to make more 4-yard than 40-yard gains. The important thing is to keep the ball moving in the right direction.

Some at both ends of the political spectrum, however, see any kind of compromise as a betrayal and blindly reject bipartisan cooperation out of hand. They forget that the Constitution itself was the product of hard-fought negotiations. So were the Emancipation Proclamation and the Civil Rights Act of 1964. In fact, most of the nation's great legislative accomplishments, each a new model in its day, occurred because a majority was willing to negotiate with the minority in good faith for the good of the country and vice versa.

Congressman Wes Watkins, my colleague from Oklahoma, has reminded me several times that there is a difference between compromising a principle and the principle of compromise. Today, our nation is as evenly divided politically as it has ever been in our history. Neither side has the political power to move its agenda forward without the cooperation of the opposition. One can argue the merits of this kind of divided government, but the reality is that bipartisan cooperation is the only way to make any progress in solving the problems facing our country.

Yet many on the far right and far left refuse to accept the current political facts of life. When it comes to moving legislation, it's their way or no way. Building coalitions to pass legislation may require us to move more slowly than some of us would like and occasionally to swallow hard, but as long as we are moving in the right direction, as long as our basic principles are not compromised, some progress is better than none. And the American people expect us to act. They don't see problems in ideological terms as we do in Washington. They don't see the Education Reform Bill, which was passed last year, as a compromise but as a solution. They don't worry whether the final bill signed by President Bush was a conservative or liberal victory. They judge its merits by one simple question: Will it help my children get a better education?

If we follow these strategies to make America a better place and put aside partisanship when it becomes a destructive force, we can make America the "shining city on a Hill" that President Reagan spoke of so often and so eloquently.

There's a wonderful story I've often told of a little boy and his father. The father is trying to get some work done, and his son wanted some attention. Sounds familiar. To keep him occupied, Dad gives the little boy a marker and some paper. "Draw a picture of the family," he told him. Two minutes later, the boy was back proudly showing off the stick figures he'd drawn.

But Dad needed to buy some time.

"Go draw the dog," he ordered. Back comes Junior with a stick figure of the dog. "This strategy isn't working," the father thought. Then he spotted a magazine lying on the floor; and as he glanced at it, he saw a picture of the world.

In what he thought was a stroke of genius, he tore the picture from the magazine and ripped it into twenty pieces.

"Here, son," he said, "go put the world back together."

It wasn't but five minutes later that the boy returned, saying, "Look, Daddy, look what I've done!" The father looked and said, "Son, how did you do that so fast?"

The boy answered, "It was easy. There was a picture of a man on the back, and once I put him all together, the world fell into place."

God put us on this earth, in part, to prepare the way for those who will come behind, and that's what all of us must do. The old ways of doing things are tired ways that no longer accomplish their mission—if they ever did. New models of innovative partnerships between government, social organizations, churches, youth groups, and a hundred others is the future I see if we have the will to change. And if we have the will to change, then we will change lives for the better.

CONCLUSION

It is a dreadful misfortune for a man to grow to feel
that his whole livelihood and whole happiness depend
on staying in office.

Theodore Roosevelt

Back in my college days, the "glory" years, a friend of mine, Jerry D. Brown, got me hooked on wearing boots. Twenty-one years later, it's a habit I still haven't kicked. Luckily for me, we now have a president who's made boots fashionable again even with a tux.

Back then, however, my only fashion statement was my infamous Afro. So, Jerry also took me around to his favorite barbershop, which I never used for a haircut but I did love the conversation. It was there I met a man by the name of Warren, who shined shoes, and soon I was a regular customer, getting my shoes shined to a fare-thee-well for the bargain price of $2. Just a couple of weeks after I had come back from the Orange Bowl, I was still strutting around Norman and Oklahoma City, reveling in the team's win and being named MVP. The attention was pretty heady stuff for a 21-year-old. On this particular day, I was feeling pretty good about myself and what I'd accomplished when I arrived at the barbershop for a quick shine.

Warren and I were shooting the breeze as he did my shoes, and we soon found out that we had a lot in common. I also found out that Warren had sent three kids all the way through college by shining shoes.

Although he didn't mean to do it, Warren's conversation put me in my place that day. I had had my moment of celebrity. The reporters and photographers, the pro scouts in the stands, the cheering fans—it was great and I enjoyed every minute of it, but as I talked with this dedicated and loving father, I realized that Warren was the real hero. Weighed against what's really important in life, what he had accomplished made winning the Orange Bowl seem insignificant.

Life in Washington is a little like winning the big game. It's easy to be led astray by all the hype and attention. It's easy to forget who sent you there in the first place and why. Most of all, it's hard to let go when the game is over.

I've loved being a congressman, serving the people of my district and, hopefully, the interests of my country. I've learned there is so much good in the worst of us and so much bad in the best of us. And now, again, I've put my trust in God that he will bless me with the wisdom of knowing where to go next. He's always sent me down the right path, and he will again.

In writing this book, I have spent a lot of time reflecting on my life, my accomplishments, my family, and my roots. It's been a little like opening an old scrapbook full of memories and taking a trip back to those moments that stick with you.

I remember first grade, when I first saw the little girl I would fall in love with and who would become my wife so many years later.

I remember Mama sitting in the light of an old lamp darning socks as she waited for us to finish dinner. She would get leftovers.

I remember seeing my father cry for the first time on the day Martin Luther King, Jr., was killed, and I remember the gun he carried for weeks afterward.

I also remember seeing the grieving figure of Coretta Scott King lean over and give one of her sobbing children a hug as they sat together in a pew at her husband's funeral.

I remember the first time I saw Lawrence Taylor head in my direction and I was holding the ball.

I remember the faces of my children on the day they were born and the Bible that Frankie held as I took my oath of office to become a member of Congress.

And finally, I remember the "door of no return," through which my ancestors passed and made all in my life possible.

I have said before I am a man of faith and hope. God, my parents, and Frankie gave me those gifts; and when I look through my imaginary scrapbook, I realize my life is defined by the people in it. I see photos in my mind of Mama and Daddy helping a neighbor in need, and Uncle Wade agitating for civil rights. I see again the brave fireman carrying a small child from the rubble of the Murrah Federal Building, and I see the heartbreak of another tragic disaster that tested the faith and hope of all of us in Oklahoma City—the tornado of 1999.

I have lived in Oklahoma all of my life, and I had never seen or even been close to a tornado before May 3 of that year. My youngest son, Trey, had a Little League baseball game late that afternoon. I had to leave early because I had a 7 o'clock flight to catch that night. It was a great game, and I hated to go, but duty called. I left the field around 5:15 to go home and clean up before heading to the airport. The sky was getting darker and darker, but spring storms are usually nothing to get excited about in Oklahoma. By the time I got home 15 minutes later, the rest of the game had been canceled, and the sky had turned an ominous black. I decided to stick close and waited anxiously for Frankie, who by then was rounding up all the kids to get them home. Television was mostly static, so we listened to the radio.

From time to time, I would go outside to stand and look around. The sounds of whistles and horns and sirens were coming at me from all directions. I remember thinking the evening had a doomsday feel to it. Had I not seen how dark and foreboding the skies had become, I would never have believed a description of it. I'd never seen a sky quite like it, and I hope I never do again.

I was certainly no expert on what to do in a tornado. When I was growing up in Eufaula, bad weather forced us down into the cellar a few times and that was always a frightening time for a child. But for the most part, Daddy just stayed there in the house and rode it out. That's what we—Frankie and the kids and I—decided to do.

I'll never know how we would have fared if the tornado had veered any closer to our home. But even so, I'll never take that kind of chance again. The next storm of this magnitude will find the Watts clan hunkering down in a shelter. The tornado came within 4 or 5 miles of our home and hit very hard just a mile or two farther. The houses in its path looked as though someone had taken them apart, board by board, and just scattered them around the neighborhood. I saw cars stacked on top of each other, four and five deep, piles of boards 12 and 14 feet high, even a dead horse wedged between two parked cars.

I saw entire neighborhoods devastated, totally wiped out. It could have been much worse, but fortunately most people heeded the warnings and got to safety. Thanks to local weather forecasters and a state-of-the-art radar installation at the Norman Airport, a joint effort between the University of Oklahoma and the National Oceanic and Atmospheric Administration, residents in the area got 30 to 40 minutes' warning—and it saved many lives.

But nothing could save these neighborhoods from the power of a killer tornado of this magnitude. It was simply devastating. A pastor friend of mine opened his church to be used as an emergency shelter by the Red Cross, Salvation Army, and others. I got there about 6:45 the next morning. The choir loft had been converted into a makeshift morgue.

The governor and I and several others drove through the storm-hit area to assess the extent of the damage. None of us expected what we saw and heard. One fellow we came across was standing by the road with a clothes basket that held a few items. He told us, "This is all I've got to start over, but at least I've got my life."

I heard people say words to that effect over and over again in the next few days. Most were just grateful to have survived. Emotionally, though, it was like the aftermath of the Oklahoma City bombing. Reporters asked, "What do you say to these people?" What *can* you say?

Sometimes we have circumstances in our lives that are so devastating that there isn't a whole lot you can do or say to make things better. You can only let people know that you're there to hurt with them; to cry with them. You're just there to help and be their friend.

As people waited to get back to their homes, I waited with them and saw their resilience and strength and something else—faith in their God, their community, and in themselves. My faith always encourages me in those times. It's natural to think, "Where is God? Why did this happen?"

I believe God understands those feelings. My faith always teaches me that sometimes things happen that we don't understand. But my faith encourages me to believe that in such times, our prayer just needs to be, "Dear God, I can't see your hand in this, but I trust your heart. I don't understand this, but I trust you."

There's a beautiful old hymn, "Farther Along," that talks about the vicissitudes of life, and offers the wisdom:

> Farther along we'll know more about it,
>
> Farther along we'll understand why
>
> Cheer up my brothers, live in the sunlight
>
> We'll understand it all by and by.

As I watched these fine people pick themselves up and rebuild their lives, it reminded me of something written by Marianne Williams: "In every community, there is work to be done. In every nation, there are wounds to heal. In every heart, there is the power to do it."

As I finished this book, one last snapshot came to my mind's eye. In one of the worst-hit neighborhoods, I saw what was left of someone's home—a pile of glass and boards and brick, the family possessions of a lifetime strewn about the yard like so many dead leaves. But for me it symbolized all that is good and great about this country and its people because there atop all the debris and rubble an American flag proudly waved.

It was a defiant statement: "This building is gone, but we're still standing."

I've never been prouder to be an American nor have I ever felt more privileged to serve the people of Oklahoma's Fourth District than at that moment.

I don't know where I'll be 5 years from now or what I'll be doing. I know there are still some things I'd like to accomplish in my life. But if God took me tomorrow, I'd go with a smile on my face, serenity in my heart, and a lifetime of memories that most people only dream of. And for that I am grateful.

ACKNOWLEDGMENTS

When one has been as blessed as I have with family, friends, mentors, and colleagues, it's difficult to know where to begin and how to end in saying thanks. So I'll begin at the beginning with the family that has been my rock through good times and bad. My parents and brothers and sisters, my aunts and uncles and cousins, and now my loving wife, Frankie, and all my wonderful children have given my life meaning and given me the most important thing in life—their love.

Mere words seem inadequate to express the deep gratitude, admiration, and respect I feel for those brave men and women of the Civil Rights Movement who risked their lives so that my generation and generations of black children to come might have an equal opportunity to reach for the stars. To Dr. Martin Luther King, Jr., Congressman John Lewis, Rosa Parks, my uncle Wade, and my father, I want to give special thanks from my heart.

I'm grateful to all the coaches I've had the pleasure of learning from, especially Coaches Paul Bell, Perry Anderson, and Barry Switzer. Each has had a tremendous impact on my life. So did all my teachers along the way, from elementary school through high school. They opened doors for me and helped me believe in myself and the possibility of my dreams.

I'm also grateful to the community of Eufaula for the roots of my raising, for keeping me safe and out of trouble most of the time, and for teaching me the values of small-town life.

To Pat Adams and Susie and Gary Moores, you've always been there for me, and I consider myself a lucky man to call you friends. Thank you for everything.

And my thanks, too, to all the Sunnylane Baptist Church youth volunteers for loving our Sunnylane kids enough to get involved.

Over the past 8 years, my congressional staff has become like a second family to me. I am proud of each and every one of them for the commitment and compassion they have shown for the people of Oklahoma and grateful for their loyalty and love of public service. I especially want to thank my first chief of staff, Mike Hunter, who helped put it all together, and my current chief of staff, Pam Pryor, who keeps us all on track today.

I also want to thank all the members of my campaign team over the years, especially Chad Alexander, who loves political warfare, and Tom Cole and Jeff Cloud, who both have great political instincts and love the political trenches. And of course Jodie Thomas and Tim Crawford, who handle my Washington political shop with integrity and professionalism.

These acknowledgments wouldn't be complete without a nod to Jack Kemp. It was Jack Kemp's challenge to the Republican Party to help underserved communities that connected with me so many years ago. His moving words drew me into the party and were the "pilot light" for my later efforts to pass the Community Renewal Act.

I also want to thank my co-author, Chriss Winston, for her patience, and Byron Laursen, for his contributions to this book as well.

My thanks also go to my agent, David Vigliano, for his support and guidance throughout the process; my editor, Mauro DiPreto, of Harper Collins, for his professionalism in getting us over the speed bumps to publication; and to his assistant editor, Joelle Yudin—no book ever had a better mother hen.

Finally, my deepest thanks go to the good people of Oklahoma, who have supported my efforts on both the gridiron and in the political arena. I've had so many wonderful experiences over last 12 years—thank you for giving me the opportunity to serve.

INDEX

abortion, 205, 219, 267
Adams, Pat, 90, 217, 277
Adenauer, Konrad, 52
AFDC, 194
affirmative action, xix, 205–9,
    251, 252, 260
Africa, 256–58
African Americans, 179
    animosity toward Republican
        party among, x, xii–xv, 150,
        253–54
    black conservatives criticized
        by, 199–200
    conservative, xii, xviii–xix, 2–3,
        21, 236, 246, 248–49
    cultural values of, 151
    in Democratic Party, 148–51,
        159–60, 186, 213, 246–47,
        250–51
    depictions of, 52–53
    economic mobility among, 2
    education and, 2, 63, 68–70
    group identity among, 2–3,
        203, 249–50, 252–53
    Jim Crow laws and, 2, 14, 35,
        54–55, 57, 176, 254

liberal, 2
    lynching of, 54
    as political independents,
        150
    political power of, 176
    in Republican Party, 1–2, 117,
        150, 154, 186–87, 207,
        216–17, 249–50, 253–54
    role models for, 63
    school desegregation and, 40,
        53–54, 57–60, 75, 84, 254
    self-affirmation vs. victimhood
        in, 34–35, 69–70
agriculture, 267
AIDS, 245
Air Force, 162
alcohol, 64
Alexander, Chad, 278
Ali, Muhammad, 118
allowances, for children, 43–44
Al-Qaeda, 259
Altus Air Force base, 179
American Community
    Renewal Act, 195, 216,
        260–65, 278
Anderson, Bill, 170, 212

8/03

B
Watts
Watts
What color is a conservative?